A-LEVEL
STUDENT GUIDE

AQA

Geography

Human geography

David Redfern
Catherine Owen

HODDER
EDUCATION
AN HACHETTE UK COMPANY

Photo credits

p.145: Yvonne Follows-Smith and David Redfern; p.146: David Redfern

Acknowledgements

Every effort has been made to trace all copyright holders, but if any have been inadvertently overlooked, the Publishers will be pleased to make the necessary arrangements at the first opportunity.

Although every effort has been made to ensure that website addresses are correct at time of going to press, Hodder Education cannot be held responsible for the content of any website mentioned in this book. It is sometimes possible to find a relocated web page by typing in the address of the home page for a website in the URL window of your browser.

Hachette UK's policy is to use papers that are natural, renewable and recyclable products and made from wood grown in well-managed forests and other controlled sources. The logging and manufacturing processes are expected to conform to the environmental regulations of the country of origin.

Orders: please contact Hachette UK Distribution, Hely Hutchinson Centre, Milton Road, Didcot, Oxfordshire, OX11 7HH. Telephone: +44 (0)1235 827827. Email education@hachette. co.uk. Lines are open from 9 a.m. to 5 p.m., Monday to Friday. You can also order through our website: www.hoddereducation.co.uk

ISBN: 978 1 3983 2819 8

Contents

Content Guidance

Questions & Answers

■Getting the most from this book

Exam tips

Advice on key points in the text to help you learn and recall content, avoid pitfalls, and polish your exam technique in order to boost your grade.

Knowledge check

Rapid-fire questions throughout the Content Guidance section to check your understanding.

Knowledge check answers

1 Turn to the back of the book for the Knowledge check answers.

Summaries

■ Each core topic is rounded off by a bullet-list summary for quick-check reference of what you need to know.

Exam-style questions

Commentary on the questions

Tips on what you need to do to gain full marks.

Sample student answers

Practise the questions, then look at the student answers that follow.

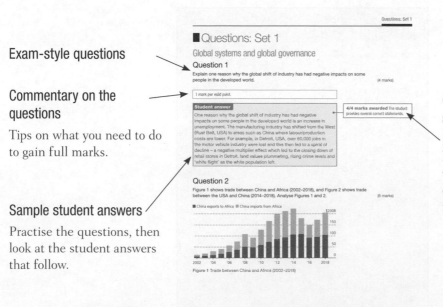

Commentary on sample student answers

Read the comments showing how many marks each answer would be awarded in the exam and exactly where marks are gained or lost.

■About this book

Much of the knowledge and understanding needed for A-level geography builds on what you have learnt for GCSE geography, but with an added focus on key geographical concepts and depth of knowledge and understanding of content. This guide offers advice for the effective revision of **Human Geography (Core and Options)**, which all students need to complete.

In the A-level Paper 2 external exam, Section A tests your knowledge and application of aspects of Global systems and global governance, and Section B tests Changing places. Section C tests one of three options: Contemporary urban environments, Population and the environment or Resource security. The whole exam lasts 2 hours and 30 minutes, and the unit makes up 40% of the A-level award. More information on the exam papers is given in the Questions & Answers section (pages 135–39).

To be successful in this unit you have to understand:

■ the key ideas of the content
■ the nature of the assessment material – by reviewing and practising sample structured questions
■ how to achieve a high level of performance within them

This guide has two sections:

Content Guidance – this summarises some of the key information that you need to know to be able to answer the examination questions with a high degree of accuracy and depth. In particular, the meaning of key terms is made clear and some attention is paid to providing details of case study material to help to meet the spatial context requirement within the specification.

Questions & Answers – this includes some sample questions similar in style to those you might expect in the exam. There are some sample student responses to these questions as well as detailed analysis, which will give further guidance in relation to what exam markers are looking for to award top marks.

The best way to use this book is to read through the relevant topic area first before practising the questions. Only refer to the answers and examiner comments after you have attempted the questions.

Content Guidance

This section outlines the following areas of the AQA A-level geography specifications:

- Global systems and global governance
- Changing places
- Contemporary urban environments
- Population and the environment
- Resource security

Read through the chosen topic area(s) before attempting questions from the Questions & Answers section.

■ Global systems and global governance

Globalisation

Dimensions of globalisation

Globalisation has a range of meanings. The way it is defined in any context tends to reflect the priorities or prejudices of whoever is writing. Economists, historians and geographers may all define globalisation differently, according to their own interests. Some people may take a scientific viewpoint and define globalisation in a way that is free of any values or perspectives. Others believe that the impact of globalisation on people's lives means there should always be a moral or ethical dimension to its discussion.

Geographers describe globalisation in general terms as the process by which places and environments become more:

- interconnected
- interdependent
- deeply connected
- connected together in real time as part of a shrinking world (also known as time/space compression)

This 'shrinking world' concept has been enhanced by the use of information and communications technology (ICT). An important outcome of the relationship between ICT and globalisation is the growth of extensive networks of people and places. All kinds of global networks can be modelled, ranging from social networks such as Facebook and Instagram to the complex supply chains of global corporations.

ICT enables all these interactions by providing affordable, instantaneous connectivity. Over time, network connections have become faster and more inclusive of all people in society, not just privileged groups. As a result, a staggering amount of globally available digital information has been created since 2000.

Globalisation The growing economic interdependence of countries worldwide through the increasing volume and variety of cross-border transactions in goods and services, freer international flows of capital and labour, and more rapid and widespread diffusion of technology.

Knowledge check 1

Illustrate some aspects of the 'shrinking world'.

Types of globalisation

Economic globalisation

The growth of transnational corporations (TNCs) has accelerated cross-border exchanges of raw materials, components, finished manufactured goods, shares, portfolio investment and purchasing. ICT supports the growth of complex spatial divisions of labour for firms and a more international economy. The internet has allowed extensive networks of consumption to develop (such as online purchasing on eBay and Amazon). Consequently, many TNCs have a global presence and a global brand.

Political globalisation

The growth of trading blocs (e.g. the European Union (EU) or the North American Free Trade Agreement (NAFTA), now replaced by the United States–Mexico–Canada Agreement (USMCA)) allows TNCs to merge and make acquisitions of firms in neighbouring countries, while reduced trade restrictions and tariffs help internal markets to grow.

The G7/G8 and G20 groups of countries meet regularly to discuss global concerns such as the global economy and the environment (the financial crash of 2007 and climate change being two major issues of recent years). The World Bank, the International Monetary Fund (IMF) and the World Trade Organization (WTO) work internationally to harmonise national economies. Note these are dominated by western democracies which all spread the view that capitalist (and consumerist) societies are the most 'successful'.

Social globalisation

International migration has created extensive family networks that cross national borders. City societies across the globe have become multi-ethnic and pluralistic. Global improvements in education and health can be seen over time, with rising life expectancy and literacy levels, although the changes are by no means uniform or universal. Social interconnectivity has grown over time due to the spread of universal connections such as mobile phones, email and the use of social media. International tourism is a major form of social globalisation, but as the coronavirus crisis in 2020 showed, it is an unstable activity.

Cultural globalisation

Successful Western cultural traits have come to dominate in some nations – sometimes called the Americanisation (or 'McDonaldisation') of tastes and fashions. 'Glocalisation' is a more complex outcome that takes place as old local cultures merge and meld with globalising influences. The circulation of ideas and information has accelerated due to 24-hour news reporting; people also keep in touch using virtual spaces such as Facebook, Skype, Twitter and WhatsApp. Due to its encrypted nature, the latter has been linked to the globalisation of terrorism.

Environmental globalisation

This refers to the increasing awareness of global environmental concerns, such as climate change and threats to Antarctica, and the need to address them. Key protagonists, such as Greta Thunberg and Extinction Rebellion, have played a major role.

Making connections

The greater connectivity (such as through social media) facilitated by the internet has been used to great effect in the management of hazards – for example, alerting people, and raising emergency funding (see AQA A-level Geography Student Guide 1).

Exam tip

The use of the internet and its impact on globalisation is changing all the time. Be prepared to be very topical in this area.

Exam tip

New trade deals are often being negotiated. Monitor these, for example trans-Pacific trade arrangements.

Globalisation can also be said to be the continuation of a far older, and ongoing, economic and political project of empire-building. It can therefore be seen as a series of periods over recent times:

- **Period 1 During the 1950s:** the end of the age of colonial empires with, for example, independence for African and Asian countries; postwar settlements and bodies (IMF, World Bank, General Agreement on Tariffs and Trade (GATT) and the WTO); the growth of container shipping.
- **Period 2 Between the 1960s and the 1980s:** the rise of the Asian Tigers (Hong Kong, South Korea, Taiwan, Singapore) and deindustrialisation of parts of Western core countries; the rising influence of the Organization of Petroleum Exporting Countries (OPEC) and their oil price rises; offshoring beginning in earnest; the spread of financial deregulation and the growth of money markets.
- **Period 3 The 1990s to the present day:** the fall of the Soviet Union; the growth of the BRIC (Brazil, Russia, India, China) economies; the growth of trade blocs (EU, NAFTA); the acceleration of technology; punctuated by periods of economic crisis.
- **Period 4 The future:** the rise of China and India and some new emerging economies such as Indonesia, Malaysia and Turkey.

Global marketing

When a company decides to embark on **global marketing**, it views the world as one market and creates products that fit the various regional marketplaces. The ultimate goal is to sell the same thing, the same way, everywhere – creating a global brand. Some examples include:

- **Coca-Cola:** the company uses the same formulae (one with sugar, the other with corn syrup) for all its markets. The classic contour bottle design is incorporated in every country, although the size of bottle and can that is marketed is the same size as other beverage bottles and cans in any particular country.
- **Mars:** the company introduced a chocolate bar called 'Snickers' around the world, though for a time, until 1990, it was marketed under the name 'Marathon' in the UK and Ireland.
- **Lever Brothers:** in 2000, Lever Brothers (UK) decided to rename its household cleaning product 'Jif' to 'Cif', the company's global moniker for the product.

It seems that the common factor is the need to align with markets in the USA and the far east.

Global patterns of production, distribution and consumption

Globalisation has created an international division of labour. At its simplest, it is possible to recognise two groups:

- Occupations that are highly skilled, highly paid and involve research and development (R&D), decision-making and managerial roles. These are largely concentrated in developed (or high-income) countries.
- Occupations that are unskilled and poorly paid assembly roles. These tend to be located in the emerging economies, often developing (or low-income) countries, based on their low labour costs.

Knowledge check 2

In addition to the BRIC countries, geographers now refer to the MINT countries. Name them.

Global marketing
Marketing on a worldwide scale, reconciling or taking commercial advantage of global operational differences, similarities and opportunities in order to meet global objectives.

Exam tip

What other products have you seen on your holidays that are the same or similar to those in your country? You can use these examples to support your answers.

This division has arisen from:

- many countries that were previously classified as developing becoming **NEEs**
- **FDI** by TNCs towards those NEEs which enabled manufacturing at a competitive price – a movement called the 'global shift'
- the **transfer of technology** which enabled NEEs to increase their productivity without raising their wages to the levels of developed countries

The main outcomes have been:

- by 2018 over half of manufacturing jobs were located in the NEEs and over 60% of exports from such countries to the developed world were manufactured goods
- **deindustrialisation** – though this is not entirely due to the 'global shift', as other factors such as outmoded production methods, long-established products entering the end of their life cycles and poor management all contributed

---

NEEs Newly emerging economies, such as China, India, South Korea and Indonesia.

Foreign Direct Investment (FDI) The money invested into a country by TNCs or other national governments.

Transfer of technology The movement of ideas and technology from one region or country to another.

Deindustrialisation An absolute, or relative, decline in the importance of manufacturing in the industrial economy of a country and a fall in the contribution of manufacturing to GDP.

The provision of services has become increasingly detached from the production of goods. The financial sector, for example, has no direct relationship to manufacturing. Therefore, as manufacturing has become more dispersed worldwide, high-level services have increasingly been concentrated in places different from the old centres of manufacturing.

The 1990s saw the emergence of a growing number of TNC service conglomerates, particularly in banking and other financial services, and in advertising. The movement of capital around the world has speeded up, such that many service industries are owned by TNCs, purely for their financial gain. These include private equity firms and venture capitalists.

Another recent trend has been the decentralisation of low-level services from the developed to the developing world. Call centre operations, for example, have moved from the UK to India, the middle east and South Africa, where employment costs are generally at least 10–20% lower. This globalisation of services is simply following what has been happening in manufacturing over several decades.

In terms of consumption, as NEEs have developed, their populations are becoming more affluent and demanding similar consumer products and services to the developed world. This may mean that trade becomes increasingly focused towards east Asia, and intra-Asian trade will also increase. Consumption of financial services is also increasing in Asia, leading to western TNCs expanding there. This is mirrored by the growth of 'branches' of western universities and British independent schools in this region.

Knowledge check 3

Where are the major financial services located in the world today?

Factors in globalisation

The development of technologies

Technological developments that have taken place to create a **shrinking world** to facilitate globalisation include:

- the development of containers – intermodal metal boxes that can be carried by ship, lorry or train
- the growth of the logistics industry to manage the movement of containers
- developments in ICT and mobile technology that enable **time/space compression** – satellites, fast broadband, networks of fibre-optic cables, smartphones
- the use of these technologies by businesses to keep in touch with all elements of production, supply and sales, and to transfer money and investments
- the use of these technologies by individuals for social networking, banking, shopping and leisure activities
- the use of these technologies to create a range of security-based systems and industries

As a consequence of technological developments, security for container transport, air transport and financial transactions, and cybersecurity, are now major issues for the globalised world

Management and information systems

TNCs have modified their management and information systems. These include:

- **global production networks (GPNs)** – large corporations (ranging from Dell to Tesco) have established multiple subcontracting partnerships while building their global businesses, often coordinated by a **hub company**
- **just-in-time production** – this requires a very efficient ordering system and reliability of delivery. Production is said to be 'pulled through' rather than 'pushed through'
- 'zero defect' – TNCs have very strong links with their suppliers, which are monitored rigorously

The management of all of the above can be undertaken remotely using ICT.

Global production network (GPN) A system whereby a TNC manages a series of suppliers and subcontracted partnerships while building its global business.

Hub company A company that orchestrates production on a global scale.

Just-in-time (JIT) production A management system that is designed to minimise the costs of holding stocks of raw materials and components by carefully planned scheduling and flow of resources through the production process.

Trade agreements

As stated earlier, most governments actively seek global connections in the belief that trade promotes economic development and wealth, by:

- joining **free trade blocs** such as the European Union (EU) and Association of South East Asian Nations (ASEAN), which make trade barrier-free between member states, and in the case of the EU allows free movement of people between countries
- opening up their markets to competition and enabling **free market liberalisation**

Shrinking world The idea that the world feels smaller over time, because places are closer in terms of travel or contact time.

Time/space compression The idea that the cost of communicating over distance has fallen rapidly.

Free trade bloc An agreement between a group of countries to remove all barriers to trade such as import/export taxes, tariffs and quotas.

Free market liberalisation Ending monopoly provision of some industries and services like telephone, gas and electricity, so people can choose a supplier based on quality and price.

Global systems

This section will examine the form and nature of the interdependencies and interconnections (economic, political, social and environmental) in the world today.

Globalisation and global systems have created a number of issues including:

■ unequal flows of people, money, ideas and technology, which in some cases can act to promote stability, economic growth and development but can also cause further inequalities, conflicts and injustices for people and the places where they live

■ unequal power relations that enable some nations to drive global systems to their own advantage and directly influence geopolitical events, while other nations are only able to respond or resist in a more constrained way

Interdependence

Some key points:

■ Economic **interdependence** through trade can be exemplified by a developed country exporting manufactured goods to a developing country, and importing raw materials in return.

■ Socioeconomic interdependence occurs where economic migrants provide labour in a country and on their return bring newly acquired skills, ideas and values to their home country.

■ Increasingly countries are becoming interdependent in their effects on the global environment and in their political relationships.

■ **Global governance** is struggling to keep up with the pace and extent of the economic and social interdependence of capital and trade flows, illegal and legal migration of people and technological change.

Interdependence The mutual dependence of two or more countries in which there is a reciprocal relationship.

Global governance The norms, rules and institutions that regulate trade and other global systems.

Inequality of flows

Unequal flows of people, money, ideas and technology within global systems can:

■ sometimes act to promote stability, growth and development

■ cause inequalities, conflicts and injustices for people and places

Stability, growth and development

Stability

Some key points:

■ Trade contributes to international peace and stability, especially if countries trade under the same rules, such as those of the **WTO**

■ Trade encourages states to cooperate – multilateral and bilateral trade agreements contribute to economic and political stability.

Exam tip

Free trade and 'Fairtrade' are often confused. Make sure you understand the difference.

Knowledge check 4

What is just-in-time production?

Making connections

A systems approach is a key concept of A-level geography. It can apply to human geography as much as it does within physical geography (see AQA A-Level Geography Student Guide 1).

WTO World Trade Organization.

■ Some bilateral agreements extend beyond trade, and may lead to cooperation and assistance in dealing with political issues such as strengthening democratic processes and human rights (such as combating child labour).

■ All of the above create a more stable environment for foreign investors.

Economic growth

Some key points:

■ Trade in merchandise and commercial services stimulates production, contributes to GDP growth and to further investment, including FDI.

■ Employment opportunities are created, incomes are raised and in some developing countries poverty levels can be significantly reduced.

■ The **economic multiplier** can be enhanced by international trade.

■ Foreign exchange (monetary flow) generated by trade can stimulate further domestic and foreign investment.

Development

Some key points:

■ Removal of tariffs and other obstacles to trade helps to generate foreign exchange which can be invested to reduce internal inequalities in poverty, health, education, infrastructure and transport.

■ The **corporate social responsibility** of TNCs can be of economic and social benefit to employees and communities in areas of production.

■ Membership of regional trading blocs and political unions can help socioeconomic development within member states.

■ Migration of highly skilled workers (scientists, engineers) can be innovative in circulating ideas and information on technology development between countries.

Inequality, conflict and injustice

Inequality

Some key points:

■ Many developing countries have limited access to global markets – this widens the **development gap**.

■ Skilled workers, especially men, tend to benefit most from employment opportunities created by trade, whereas many unskilled workers and women held back by limited educational opportunities remain unemployed and unable to contribute to the workforce.

■ In developing countries, internal inequalities are exacerbated by trade activity, often spatially concentrated in ports, where most commercial activity is located.

■ It leads to internal migrant flows and widening inequalities within the country.

Conflict

Some key points:

■ Disputes can arise over tariffs, prices of commodities and changes to trade agreements.

■ Border and customs authorities can be subject to corruption and breaches of security.

■ Port development, mining and deforestation create environmental conflicts.

Economic multiplier
An initial investment in an economic activity in an area has beneficial knock-on effects elsewhere in the area's economy.

Corporate social responsibility
A TNC's commitment to assess and take responsibility for its social and environmental impact – including its ethical behaviour towards the quality of life of its workforce, their families, and local communities.

Development gap
The difference in prosperity and wellbeing between rich and poor countries – measured by GDP per capita.

Injustice

Some key points:

- Displacement of communities takes place due to land-grabbing following investments in mining or agriculture.
- Attempts to secure cheap labour can lead to child labour and forms of modern slavery.
- Unfair trade rules can adversely affect businesses such as small-scale farmers or fishermen.

Inequality of power relations

Unequal power relations mean that:

- some states are able to drive global systems to their own advantage and to directly influence geopolitical events
- others are only able to respond or resist in a more constrained way

States that drive global systems

This can be illustrated with reference to a developed economy, such as China or the USA, and how it has a strong geopolitical influence and drives the global system to its own advantage as a result of some or all of the factors in Table 1.

See the case study on China below (page 14).

Table 1 Factors that help a state drive global systems and challenges it may face

Factor	Features
The components of its international trade	■ its patterns of trade, trade partners and trade agreements ■ its membership of trading blocs
Its advantages for international trade	■ investment in domestic transport and communications infrastructure ■ industrial productivity ■ outward FDI ■ ability to exploit its own natural resources ■ political strength in negotiating trade agreements ■ levels of skill and education in the workforce
Opportunities that international trade creates for the country	■ employment in a wide range of industrial sectors ■ stimulation of the economic multiplier effect ■ development of positive political and cultural relationships with its trade partners, including stewardship of the environment ■ ability to integrate other countries, rich and poor, into their supply chains
Challenges which arise as a result of its influence	■ pollution issues and land-use conflicts, e.g. resulting from port development ■ trade disputes over 'price dumping' ■ managing a trade surplus or deficit ■ managing migration across its borders

States that respond to global systems

This can be illustrated with reference to a developing economy, such as a sub-Saharan country, and how it has a limited influence on, and can only respond to, the global system as a result of some of the factors given in Table 2.

Exam tip

Questions on this area are likely to ask for evaluation or assessment. Make sure your answer is discursive with a clear conclusion.

Exam tip

When discussing this topic, try to write about a range of economic, social and environmental aspects.

Table 2 Factors that influence a state having to respond to global systems

Factor	Features
The components of its international trade	■ its patterns of trade, trade partners and trade agreements ■ its membership of trading blocs
Its limited access to global markets	■ its limited ability to exploit, transport, market and export its primary products ■ its inability to cope with economic shock, such as changes in global demand and prices for primary products ■ its vulnerability to natural hazards, and the effects of political shock, such as conflict, on its economy
Opportunities or otherwise brought by international trade	■ economic development and diversification of industry ■ need for investment in key infrastructure ■ socioeconomic development through investment in health and education ■ human rights issues, crime and conflict ■ interdependence with, for example, the EU and USA, and/or within the same region, e.g. the East African Community
Challenges which remain	■ achieving political stability and democracy ■ removing barriers that prevent integration into global systems, such as illegal practices or limited investment in transport infrastructure ■ managing environmental problems arising from mining and forestry operations ■ reducing socioeconomic inequalities

Making connections

Vulnerability to natural hazards (e.g. earthquake, tropical storm) may limit a country's access to global markets (see Hazards in AQA A-Level Geography Student Guide 1).

Exam tip

Be aware of unequal power relations within the global system – identify some of the world's most powerful and least powerful countries.

China: case study

Unequal power relations can be discussed in the context of China and its relations with the rest of the world.

At the end of the 1970s, Deng Xiaoping's China began to 'open its doors' to the rest of the world, turning its back on the period under Chairman Mao. Xiaoping recognised that if China as a whole was to become prosperous, some regions would have to become rich before others. He allowed the coastal regions to develop at a much faster pace than the inland regions, even if inequality occurred as a result. The liberalisation of trade through joining the WTO in 2001 has also helped greatly.

A key aspect of China's growth has been the setting up of Special Economic Zones (SEZs) along the east coast. These are designated areas of very fast economic development. They have special regulations and fewer restrictions on growth, while also offering higher wages than in the rest of China. China's eastern region contains its top three municipalities – Beijing, Shanghai and Tianjin. It also includes the country's largest city economies of Guangdong Province (including Guangzhou and Shenzen), Shenyang and Hangzhou. Guangdong Province accounts for 25% of China's international trade and has become a magnet for migrant workers from across China. Its growth and economic development have been huge. However, there has been an associated increase in income inequality within the country.

In terms of industry, the state-owned enterprises (SOEs) were reformed. A key incentive was to allow them to keep some of their profits for further investment. In addition, they improved their management strategies, and some smaller SOEs were privatised. There have also been fewer barriers to collaboration with foreign partners, and hence several SOEs have attracted foreign TNCs as partners and FDI has been significant, as joint ventures (JVs). There has also been more westernisation in terms of profit-and-loss accounting, patent legislation, and scientific and technological research. Competition is now encouraged. Some major Chinese companies that have set up JVs with other foreign companies include Sinopec, PetroChina, China Mobile and SAIC Motor Corporation (formerly Shanghai Automotive).

There were other benefits from China's growth. Infrastructural improvements – ports, railways, roads, airports – built initially for industrial and trade purposes are also available for other forms of international travel or trade. They allow China to integrate more with the world and they are also a catalyst for rapidly rising living standards. It is not just the case that China can export more, it can also import more, increasing its interdependence with other countries for the goods and services which the rising numbers of middle classes desire.

Who are China's rivals?

For some time now, Brazil, Russia, India, China and South Africa have been grouped together under the acronym BRICS. The BRICS' gross domestic product (GDP) totalled US$21 trillion in 2019. This compares with US$19 trillion for the EU and US$21 trillion for the USA. However, China alone accounts for 66% of the BRICS' GDP; take China out and they are much less powerful economically.

We must also note that there are other 'kids on the bloc' such as Mexico, Nigeria, Bangladesh, Vietnam, South Korea, Indonesia, Turkey and Egypt. Many of these emerging nations share the characteristic of a **demographic dividend**. They have a window of opportunity when their workforce is large and does not have to support either a large young or old population.

China: going global?

China is expanding into the rest of the world. It has over US$2 trillion to invest overseas. China also owns a huge proportion of USA debt. So where is this investment going? Some destinations include:

- Brazil – development of its natural resources, both agricultural and mineral
- Australia – into resources of iron ore, coal and gas
- Nigeria – investment in oil, mostly through engineering and construction
- UK – buying on the property market and in shares on the London Stock Exchange
- Saudi Arabia – investment in infrastructural construction

China has recently announced a new project: the 'Belt and Road Initiative', or BRI. This project is highly ambitious and covers several aspects of global economic importance – shipping lanes, road and rail routes and pipelines. China wants to strike economic and cultural partnerships with other countries, and thereby cement its status as a dominant player in world affairs. Some examples of the project include:

- in 2016 the Chinese shipping company, Cosco, took a 67% stake in Greece's second-largest port, Piraeus, from which Chinese firms are building a high-speed rail network linking the city to Hungary and eventually Germany
- a gas pipeline from the Bay of Bengal through Myanmar to south-west China
- a rail link between Beijing and Duisburg, a transport hub in Germany

---

Exam tip

Use an atlas or online source (such as Google Earth) to locate these cities and areas so you can identify their location on a map in an exam.

Knowledge check 5

Summarise the main causes of economic growth in India.

Demographic dividend
Occurs when there are fewer dependent children and elderly people with relatively more productive adults in a population.

China is also investing heavily in sub-Saharan Africa, often in infrastructure projects. Furthermore, over 1 million Chinese people live in Africa, some for the long term or even permanently, having bought land and started businesses. Chinese trade with Africa in 2018 was US$205 billion a year, mostly exports of oil and minerals from Africa to China.

Examples of Chinese investment in Africa include:

■ **Angola:** where China has helped reconstruct the country after its long civil war. Loans, secured by access to Angolan oil reserves, have been used to build roads, railways, water systems, hospitals and schools
■ **Congo:** where Chinese technicians have built an HEP plant (to be repaid in oil) and another in Ghana (to be paid for with cocoa beans)

International trade and access to markets

The basis of the **trade** of goods and services between countries can be explained by the theory of **comparative advantage**; countries specialise in activities for which they are best equipped in terms of resources and technology. A country can then trade surpluses in order to provide the income needed to buy in goods which cannot be produced efficiently, or at all, in the home economy.

When studying these topics, you should consider how international trade and variable access to markets underlies and impacts on people's lives across the globe. In addition, you should consider how they impact on your life, and influence the way in which you live.

Global features and trends

The global pattern of international trade is uneven and complex.

■ It is dominated by developed countries (e.g. USA and Germany) and the faster-growing emerging countries (e.g. China and India).
■ In 2018, the top ten trading nations accounted for 52% of global trade in goods; developing countries accounted for 44%; the total value of exports of goods was US$19.1 trillion (Figure 1).
■ Low-income developing countries (e.g. those in sub-Saharan Africa) have limited access to international markets and as a result have relatively weak trading patterns.
■ Countries which export manufactured goods earn more from trade than those that export raw materials – manufactured goods have a higher value than raw materials.
■ Changes are taking place in global patterns of trade:
  – Emerging countries now trade more with each other than with developed countries.
  – Developed countries are exporting less to each other, and more to the emerging countries.
■ Commercial services can also be traded:
  – The largest exporters are the countries of the EU, USA and increasingly China and India.
  – There are very low provisions for the countries of sub-Saharan Africa.

Exam tip

China is also investing in the Pacific area. Investigate some of its activities in islands such as Fiji, Samoa and Vanuatu to broaden the breadth of your answers.

Trade The movement of goods and services from producers to consumers.

Comparative advantage The principle that countries can benefit from specialising in the production of goods at which they are relatively more efficient or skilled.

- In 2018, the top ten trading nations accounted for 53% of global trade in commercial services; developing countries accounted for 34%; the total value of exports of services was US$5.7 trillion (Figure 1).

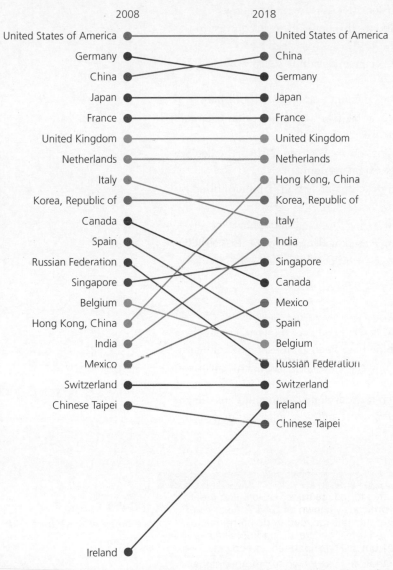

Figure 1 Changes in rank of the world's leading traders of goods and services (2018)

Source: WTO (2019)

Trends in foreign investment

In general terms:

- FDI by governments and TNCs in foreign countries is a key element of the global system – most currently emanates from developed countries.
- The emerging economies, particularly China and the Gulf states, are increasing their levels of FDI.

- Much Chinese investment is in developing countries, such as those of sub-Saharan Africa, aimed at protecting their supply lines of raw materials.

In 2019, the UN Conference on Trade and Development (UNCTAD) produced a report on global FDI trends. Its main conclusions (for 2018) were as follows.

- Global FDI flows fell by 23% to $1.43 trillion – this being in stark contrast to the growth in global GDP and trade.
- This was largely due to flows to developed economies falling sharply, by 37%, to $712 billion.
- Overall, FDI flows to developing economies remained stable at $671 billion.
- Flows to countries in Asia remained stable, at $476 billion – the region regained its position as the largest FDI recipient in the world.
- FDI flows to Africa fell, reaching $42 billion, down 21% from 2016 – the decline was concentrated in the larger commodity exporting countries.
- FDI to Latin America and the Caribbean rose 8% to reach $151 billion, lifted by that region's economic recovery.
- Flows to the least developed countries fell by 17%, to $26 billion, however.
- Flows to landlocked developing countries increased moderately, by 3%, to $23 billion.
- Small island developing states (SIDS) saw their inflows increase by 4%, to $4.1 billion.

Trading relationships and patterns

Trading relationships and patterns have been determined over the last 70 years by a series of trade agreements and principles which have been established to deal with the rapid rise of international trade and the issues associated with it. They deal with trade between the large, developed economies (e.g. the USA and the EU), emerging economies (e.g. China and India) and smaller, less developed economies such as in Latin America and southern Asia. These agreements and principles can be examined at the international and regional scale (Table 3).

Table 3 International organisations that promote trade and investment

Organisation	Commentary
World Trade Organization (WTO)	The WTO works to reduce trade barriers and create free trade – previously known as **GATT**. A series of global agreements has reduced trade barriers and increased free trade. The latest round of talks began in Doha in 2001 but still has not been agreed yet.
International Monetary Fund (IMF)	Since 1945, the IMF has worked to promote global economic and financial stability and encourage more open economies. Part of this involves encouraging developing countries to accept FDI and open up their economies to free trade. The IMF has been criticised for promoting a 'western' model of economic development that works in the interests of developed countries and their TNCs.
World Bank	The World Bank's role has been to lend money to the developing world to fund economic development and reduce poverty. It has helped developing countries develop deeper ties to the global economy, but has been criticised for having policies that put economic development before social development.

GATT General Agreement on Tariffs and Trade.

Organisation	Commentary
US–Mexico–Canada Agreement (USMCA) (formerly known as NAFTA)	USMCA's members consist of the USA, Canada and Mexico. Its main impact has been to create the maquiladora in Mexico. This is offshoring, based entirely on numerous low-cost labour forces in northern Mexico.
Association of South East Asian Nations (ASEAN)	A political and economic organisation of ten South East Asian countries – Indonesia, Malaysia, the Philippines, Singapore, Thailand, Brunei, Cambodia, Laos, Myanmar and Vietnam. Its aims include accelerating economic growth and social progress, and promoting regional peace and stability.
The European Union (EU)	The EU currently consists of 27 members: France, Germany, the Netherlands, Belgium, Luxembourg, Italy (the first six members), Denmark, Ireland (joined 1973), Greece (1981), Portugal and Spain (1986), Austria, Finland and Sweden (1995), Czech Republic, Cyprus, Estonia, Hungary, Latvia, Lithuania, Malta, Poland, Slovakia and Slovenia (2004), Romania and Bulgaria (2007) and Croatia (2013). The UK left the EU in 2020. The EU was established under the Treaty of Rome in 1957 with the objective of removing all trade barriers between member states. A major aim was the desire to form a political and economic union which would prevent the possibility of another war in Europe. Most of the countries in the EU use a single currency – the euro. The Treaty of Maastricht (1991) paved the way for monetary union which came about in 2002 with the adoption of the euro. Some countries do not take part in this zone, e.g. Denmark and Sweden.

Maquiladora
Manufacturing industries operating in a Mexican free trade zone close to the USA/Mexico border, where factories import material and equipment on a duty-free and tariff-free basis for assembly, processing or manufacturing. The products are then re-exported back to the USA and Canada.

Offshoring The manufacture or assembly of a product in a developing country using components produced in a developed country.

Knowledge check 6

Assemble three arguments for, and three arguments against, a country staying within the EU.

Exam tip

Use an atlas to locate all of the countries in these trading groups so you can describe their distribution on a map in an exam.

Differential access to markets

The trading agreements outlined in Table 3 and the various other ways in which countries become involved in the dealings of other countries, whether by trade or aid, or a combination of both (as illustrated by China's dealings overseas), have impacted the economic and societal wellbeing of the people within those nations. The global pattern of trade is one of great inequality with many developing countries, or regions of countries, having limited access to global markets.

Globally, a number of strategies seek to deal with these issues:

- **Special and Differential Trading agreements (SDTs)**, sanctioned by the WTO, exist between some developing nations and some developed countries. These encourage export diversification to reduce over-dependency on single exports.
- **Free Trade Areas (FTAs)**, e.g. the region of northern Mexico (maquiladora) (see below). During his presidency, President Trump accused this FTA of significantly undercutting US industries.
- **Special Economic Zones (SEZs)** – areas which have:
  - tariff- and quota-free zones (often ports) which allow manufactured goods to be exported at no cost
  - infrastructure such as port facilities, roads, power and water connections provided by the government, providing a subsidy for investors and lowering their costs

- very low taxes, and often there is a tax-free period of up to 10 years after a business invests
- often banned trade unions, so workers cannot strike or complain
- limited environmental regulations

You should study examples of how differential access to markets impacts economic and societal wellbeing – say, at national and individual levels.

National: Mexico and USMCA (formerly NAFTA)

A large variation of incomes has developed within Mexico as the country has modernised in the recent past, largely due to the access to the North American market that parts of the country possess as a result of NAFTA (now USMCA). For example, economic productivity in Nuevo León, a heavily industrialised state close to the American border (capital Monterrey), is at a level equivalent to that of South Korea. Here maquiladoras are common. In the south of Mexico, however, productivity is close to that of sub-Saharan Africa.

The country's industrial clusters devoted to the manufacture of cars, planes, electrical goods and equipment – categories that between them account for two-thirds of Mexico's manufacturing exports, and thus for about 20% of GDP – are largely to be found in a band next to its northern border and in the central states to the south of it. These states account for about 70% of the country's 127 million population.

Individual: microfinance schemes

For many people, microfinance and microcredit are essential and produce many benefits for poor and low-income households. Microfinance helps to unlock a community's entrepreneurial potential and allows people to access markets for their products.

Microfinance refers to a number of different financial products.

- **Microcredit:** the provision of small-scale loans to the poor, for example through credit unions.
- **Microsavings:** for example, voluntary local savings clubs provided by charities.
- **Micro-insurance:** especially for people and businesses not traditionally served by commercial insurance, this can act as a safety net to prevent people from falling back into poverty after, for example, a harvest failure or hazardous event.
- **Payment management:** for managing remittance payments sent between individuals. One of the best-known mobile solutions is M-Pesa in east Africa.

Microcredit attempts to reduce poverty and has several key characteristics.

- It often provides small loans for the working capital requirements of the rural poor, especially women.
- There is minimal risk assessment of borrowers compared with commercial banks.
- Security is rarely demanded for the money.
- Based on the loan repayment history of the members, microfinance institutions can extend larger loans to the members repeatedly.

However, this type of money lending is not without its critics. There are sometimes problems in terms of maintaining manageable interest rates, gender inequalities (experts agree that women should be the primary focus) and being able to reach those

> **Exam tip**
>
> Be prepared to evaluate the costs and benefits of SEZs and FTAs.

> **Exam tip**
>
> Study the nature and impact of the maquiladoras in Mexico in more detail, and examine the likely impact of the new trade deal known as USMCA.

in most need while still covering operational costs. There are also problems of people defaulting on their loans and the scheme losing its assets.

Transnational corporations (TNCs)

The standard definition of a **transnational corporation (TNC)** is a firm which has the power to coordinate and control operations in more than one country. However, large TNCs now dominate international trade and hence are important drivers of the global economic system. Over the last few decades, such TNCs have developed different forms and have moved into a wide range of activities.

- **Resource extraction:** particularly in the mining and oil and gas industries (for example BP, Exxon, Royal Dutch Shell and Chevron).
- **Manufacturing:** in high-end products of computers and electronics (Apple) and pharmaceuticals (GSK), large-volume consumer goods such as cars (Ford, Toyota) and tyres (Michelin), and mass-produced consumer goods with products such as cigarettes, drinks, foods, cosmetics etc. (BAT, Fosters, Unilever, Kellogg's, L'Oreal).
- **Services:** banking and insurance (HSBC), supermarkets (Walmart), advertising (Saatchi), freight transport (Norbert Dentressangle), hotel chains (Radisson) and fast-food outlets (McDonalds, KFC).

The nature and role of TNCs, and the linkages within them, have become much more complex. In particular, an important distinction has arisen between two different types of 'operation'.

- **Genuine overseas branch plant operations:** production or retailing facilities resulting from FDI and owned by the parent company. An example would be Ford, a US car company with its headquarters in the US, but having branch plants in countries such as the UK, Belgium and Mexico.
- **Business arrangements known as global production networks (GPNs):** large corporations, ranging from Dell to Tesco, have established thousands of subcontracting partnerships while building their global businesses. The term 'transnational corporation' cannot cover this. A TNC that orchestrates a GPN can be described as a hub company.

As globalisation has accelerated, the size and density of GPNs have grown. GPNs span food, manufacturing, retailing, technology and financial services. Food giant Kraft and electronics firm IBM both have 30,000 suppliers providing the ingredients and components they need and helping to generate huge revenues. The world's largest firms have multiplied the size of their supply chains many times over through corporate mergers and acquisitions (for example, Kraft acquired Cadbury's GPN in 2010, adding it to its own). The sheer size of GPNs has made tracking accountability of their actions difficult.

Other general characteristics of TNCs

Spatial organisation

- Most have headquarters and R&D in developed countries (in Europe and the USA).
- Manufacturing is usually based in areas of low labour costs, often emerging or developing countries (South East Asia and Eastern Europe).

Knowledge check 7

Write about the success or otherwise of a microfinance scheme you have studied.

Transnational corporation (TNC) A company that operates in more than one country.

Exam tip

You should reflect on the impact of TNCs on your life – how much use do you make of their products and services? This will help you illustrate your answers to evaluative questions.

Knowledge check 8

Explain how a hub company operates.

Production

- Most take advantage of **outsourcing** their production – some subcontracting arrangements can be highly complex (GPNs).
- Many TNCs also outsource their back-office and other services.

Trading and marketing patterns

There are two broad types of pattern:

- **Vertical integration:** a supply chain of a company is owned entirely by that company, from raw material to finished product.
- **Horizontal integration:** a company diversifies its operations by expansion, merger or takeover to give a broader capability at the same stage of production.

A specified TNC

You are required to study a specified TNC including its impact on those countries in which it operates. Table 4 provides general impacts.

Outsourcing A TNC subcontracts an overseas company to produce goods or services on its behalf.

Exam tip

Choose a company that you can research easily, and will interest you. Examples include Apple, Ford and Tata.

Table 4 Benefits and costs of TNCs

	For the host country	For the TNC	For the country of origin
Benefits	Generate jobs and income Bring new technology Develop new skills Create a multiplier effect	Lower costs because of cheaper land and lower wages Access to new resources and markets Fewer controls on environmental matters	Cheaper goods for sale Can specialise in aspects of production, such as R&D
Costs	Poor working conditions – sweatshops Exploitation of resources Negative impacts on the environment, small businesses and culture Repatriation of profits Possibility of corruption	Ethical issues that may impact on sales Reputational damage	Loss of manufacturing jobs Deindustrialisation Structural unemployment Political discontent

World trade in a commodity or product

You are required to study the world trade in at least one food commodity or one manufactured product. Here is an example of the latter.

World trade in cars

During 2018, global car exports amounted to over US$780 billion. Table 5 summarises the main importers and exporters of cars in 2018. It is interesting to note that several countries feature in both lists, and that they are all developed countries or emerging economies. It is fair to conclude that the bulk of world trade in cars is between rich countries where people have high disposable incomes – the USA accounts for 25% of all the world's imported cars.

At a larger scale, the EU is the second largest importer and the largest exporter of cars, with Germany by far the largest exporting country. It is worth noting that the scandal that affected Volkswagen in the autumn of 2015 did not impact on this European dominance.

Note: China and India are both significant producers of cars, but do not export large amounts. China produced over 21 million cars in 2019, almost three times the production of the second-largest producer, Japan (8 million).

Table 5 Top ten car importers/exporters by value (2018)

Rank	Importers	$ billion	Exporters	$ billion
1	USA	179	Germany	155
2	Germany	62	Japan	99
3	China	50	USA	51
4	UK	44	Mexico	49
5	Belgium	39	UK	42
6	France	38	Canada	41
7	Italy	32	South Korea	38
8	Canada	30	Spain	35
9	Spain	22	Belgium	34
10	Australia	17	France	25

Source: WTO (2019)

Globalisation critique

As globalisation continues apace, and more and more parts of the world have the desire to raise the living standards of their people to a perceived level of expectation, it is apparent that economic development cannot rise smoothly alongside social, cultural and environmental development. Some have argued that globalisation helps to integrate the world, thereby maintaining peace and a level of stability. However, while it is in everyone's interest that people live under better conditions, it is also true that the richest among us, nations as well as individuals, want to maintain their differential, even if their desire to do so impacts negatively on others.

Others believe that globalisation has created greater inequality, injustice, conflict and environmental degradation. They point to issues such as deforestation, water pollution, climate change and biodiversity loss. Some have suggested that the coronavirus pandemic of 2020 was a direct consequence of globalisation – its rapid transmission around the world being facilitated by the 'shrinking world'. On the other hand, the speedy transfer of technological solutions to the crisis around the world could only occur due to effective global communication systems.

Various anti-globalisation movements have arisen, such as Extinction Rebellion and Public Eye on Davos. Furthermore, some places in the world are still 'switched off' from globalisation due to political isolation (North Korea), physical isolation (Bhutan and Chinese Tibet) and economic isolation (the Sahel of Africa).

As China continues to expand its influence around the world, some have accused the country of engaging in a form of economic colonialism (as European countries did in the nineteenth century in many parts of the world). China strongly denies this is the case. It is, however, a development worth following. For example, the social and political events in Hong Kong during 2020 could impact globalisation significantly in the coming decade.

Making connections

There is a wide range of synoptic links to other parts of the course here – for example, the carbon cycle, climate change, and population and the environment.

Exam tip

It is important that you develop a view on these issues (supported with evidence), and consider how others may see them differently from you.

Case studies

Only one case study is specified by this area of the specification – the study of a TNC – but it may be desirable to outline some areas within the material covered above where case studies would enhance your work.

1 The impacts of named TNCs and their activities in specified areas of the world to illustrate the social, economic and environmental outcomes of their activities. Try to ensure that you take a balanced approach, citing positive benefits as well as negative costs. It is always useful to consider how different people may view these impacts, as values and attitudes are important in this area of study.

2 A case study of a specified TNC. You should be able to identify its country of origin as well as the countries where it operates or it subcontracts work to (host countries). Consider its spatial organisation – areas of headquarters, R&D and production – and study the linkages within the company. You should also seek to evaluate its marketing strategies for the products it trades.

Summary

After studying this topic, you should be able to:
- understand the factors that have driven globalisation, and the various dimensions to it
- understand the form and nature of economic, political, social and environmental dependencies that have arisen in the world as a consequence of globalisation
- recognise and evaluate issues that have arisen such as inequalities in flows of people, money, ideas, technology and in power relations
- know the main features of global trade, including the major trading relationships and patterns that exist between nations and economic groupings
- analyse the degree to which differential access to markets is associated with levels of economic development, and how it impacts on economic and societal wellbeing
- evaluate the nature and role of TNCs and their impacts on the countries where they operate
- know the world trade in at least one commodity (food or a manufactured product).
- analyse and assess the geographical consequences of global systems, in particular world trade and access to markets, and how they impact on your life and that of others
- debate the pros and cons of the impacts of globalisation by considering whether it has brought peace and stability or injustice and inequality to the world

Global governance

Global governance refers to the emergence during the last 70 years or so of norms, rules, laws and institutions that have regulated and, to some extent, reproduced the trade-orientated global systems that were discussed in the previous section, as well as other global systems (such as those involving patterns of human development and population migration). These regulations have in turn had geographical consequences for the world's citizens and the places where they live. Together, global systems and global governance have shaped relationships between individuals, states and non-state organisations (for example, the United Nations (UN), transnational companies (TNCs) or non-governmental organisations (NGOs)) around the world.

Many of these systems and laws have been responsible for positive changes in the way in which global geopolitics operates. For example, UN-sponsored agreements on human rights and genocide coupled with international law were crucial in creating the post-1945 international system following the atrocities under Nazi Germany. Another agreement is where nation states, with exclusive sovereignty over their national territories, are treated as equal partners under the auspices of the UN Charter.

Global governance has dealt with issues such as those concerning trade, security, nuclear proliferation, legality, **human rights**, sovereignty and territorial integrity, the atmosphere, laws of the sea and the protection of animals. Some of these refer to what is known as the 'global commons' – the Earth's resources that are, in theory, shared by all.

> ### Exam tip
>
> The AQA specification concentrates on trade-related matters in the sections on global systems and global governance. You could also research global governance of one other area from the list above.

There have, however, been two significant issues associated with these attempts at global governance:

- How have agencies, including the UN in the post-1945 era, worked to promote growth and stability, and yet may have also created and exacerbated inequalities and injustices?
- How have the interactions between the local, regional, national and international and global scales become fundamental to understanding the role of global governance?

The United Nations (UN)

The UN is an international organisation designed to make the enforcement of international law, security, economic development, social progress and human rights easier for countries around the world. It includes 193 countries as its member states, and its main headquarters are located in New York, USA.

The UN is the most representative intergovernmental organisation in the world today. It has made enormous positive contributions in maintaining international peace and stability, promoting cooperation among states and international development. The UN believes that only through international cooperation can humankind meet the challenges of these issues in the global and regional contexts. The UN plays a pivotal and positive role in this regard. However, some have pointed out that some of the actions of the UN have actually exacerbated inequalities and injustices.

> ### Knowledge check 9
>
> Summarise the main features of Article 1 of the UN Charter.

> **Human rights** Moral principles or norms that describe certain standards of human behaviour, and are protected as legal rights in international law.

The UN operates through applying the principles of the Charter of the United Nations and its main authority in maintaining international peace and security is through the Security Council. The UN states that to strengthen its role, it is essential to ensure to all member states the right to equal participation in international affairs, and that the rights and interests of the developing countries should be safeguarded. Although the UN does not maintain its own military, it does have peacekeeping forces which are supplied by its member states. On approval of the UN Security Council, these peacekeepers are often sent to regions where armed conflict has recently ended, in order to discourage combatants from resuming fighting.

In addition to maintaining peace, the UN aims to protect human rights and provide humanitarian assistance when needed. In 1948, it adopted the Universal Declaration of Human Rights (UDHR) as a standard for its human rights operations. The UN currently provides technical assistance in elections, helps to improve judicial structures and draft constitutions, trains human rights officials, and provides food, drinking water, shelter and other humanitarian services to people displaced by famine, war and natural disaster.

In 2000, the UN established its **Millennium Development Goals (MDGs)**. Most of its member states and various international organisations agreed to achieve these goals relating to reducing poverty and child mortality, fighting diseases and epidemics, and developing a global partnership in terms of international development by 2015. Some member states have achieved a number of the agreement's goals while others have reached none. This is cited as one example where the UN may have exacerbated inequality around the world.

The Sustainable Development Goals (SDGs) were agreed in September 2015 with some targets similar to the MDGs, such as ending poverty and hunger, and others focused more on combating the threat of climate change and protecting oceans and forests from further degradation. To be successful, the SDGs will require a renewed UN system. A growing number of emerging nations will play an expanded role in this system, with probably a larger financial contribution, greater presence in governance, a stronger voice and a greater influence.

Interactions

All producers and consumers are linked with other people in other places and at different scales (local, national and global). Their interdependence and interactions are crucial for many of the world's systems. Although the power to act and to effect change is embedded in many different locations within the system(s), the most effective changes are often brought about by different people or places working together in some form of partnership. Here is a list of some of the partners (sometimes referred to as 'actors') that can effect change at a global scale:

- **TNCs:** these can form or encourage cooperatives; they can source their materials and products ethically; they can enforce codes of conduct of their producers…or they can deliberately do none of these.
- **National governments:** these can seek to regulate TNCs and these regulations can be replicated by other countries.
- **Supranational bodies (such as the European Union (EU) and World Trade Organization (WTO)):** these can regulate trade.

Exam tip

Research where the UN currently has peacekeepers in the world.

Knowledge check 10

Through which agencies does the UN provide humanitarian assistance around the world?

Millennium Development Goals (MDGs) A series of targets between 2000 and 2015 that aimed to act on the main causes of poverty around the world including diet, education and disease.

Exam tip

You should examine the success or otherwise of one or two specific MDG targets.

- **Workers:** they can form trade unions to defend their rights, both nationally and perhaps internationally; in extreme cases they can cite solidarity internationally.
- **Consumers:** they can ask moral questions about the origin of food and other products, and they can reject exploitative goods.
- **Farmers:** they can organise themselves into collectives and have greater strength to negotiate as groups. An example of this is the Fairtrade movement (see below).
- **NGOs (such as Greenpeace, Oxfam):** these can lobby, raise public awareness, fund projects and educate.

Finally, there have been some concerns expressed about global governance of whatever is being governed, with several questions being raised:

- What is the purpose of the governing mechanisms? How and why were the particular agencies/partners brought together and what are their interests and rationales?
- How well do the various agencies/partners work together?
- How well does global governance work bearing in mind the different rates of economic, social and cultural development around the world?
- Just how democratic or accountable are these unelected, and largely appointed, bodies and the people who run them? How does this square with increasing levels of inclusion and empowerment?

The development of trade agreements

The first major steps in the governance of today's level of international trade took place after the Second World War. The initial steps in opening up the global economy were made in the 1950s and 1960s as the desire to avoid the economic and political mistakes of the interwar period led to the gradual dismantling of trade barriers. They also came about as a result of close cooperation between democratic powers keen to escape the conflicts of the first half of the twentieth century. This resulted in:

- the Marshall Plan, which helped to deliver postwar reconstruction in a bid to avoid the mistakes contained within the Treaty of Versailles
- the delivery of a number of successful General Agreement on Tariffs and Trade (GATT) rounds designed to reduce trade barriers. GATT was replaced by the World Trade Organization (WTO) in 1995
- the creation of the International Monetary Fund (IMF) and the World Bank (see also Table 3)
- the creation of the North Atlantic Treaty Organization (NATO), a response to fears of Soviet expansion in Europe (an example of global security governance being interdependent with trade governance)

These collectively interconnected a series of systems and arrangements which were political, financial, economic and security-based in order to support trade. The consequence was a massive reopening of world trade. However, only those countries which make up the core countries within the developed world, the **OECD**, really benefited.

As the reforms took place between like-minded democracies, it seemed as though economic progress depended on democratic status. There was one major exception to this, which remains so to this day – Singapore. Singapore's economy has gone from

Exam tip

A consistent theme of human geography at A-level is how a concept impacts on your life and that of others across the globe. You need to develop views and opinions, and the confidence to express them.

Knowledge check 11

What is the purpose of NATO?

Organisation for Economic Cooperation and Development (OECD) A group of economically developed countries that aims to promote policies which will improve the economic and social wellbeing of people around the world.

extreme poverty through to considerable wealth in the space of 50 years even though it has no conventional democratic framework.

During the latter half of the twentieth century, more and more countries began to not only trade with, but also, through TNCs and government sources, invest in other countries (FDI). In the initial stages, the rapid rise of FDI mostly affected wealthy (high-income) nations. Japan and Germany, with their large current account surpluses mostly lent money to the USA which ran a large current account deficit. In return for this funding of the USA's deficit, the latter provided military and diplomatic protection for both countries in the midst of the Cold War. This is another example of where trade and security come together.

At the end of the 1970s, Deng Xiaoping's China was beginning to 'open its doors' to the rest of the world and turning its back on the period under Chairman Mao. Similarly, India, which had previously protected inefficient domestic industries, began to reconnect with the West. Countries in Eastern Europe previously controlled by Soviet Russia rushed to join the EU. Countries in Latin America also wanted to join in the spirit of free trade – Mexico joined the North American Free Trade Agreement (NAFTA,) and several Latin American nations formed Mercosur, another free trade union.

All of this is ultimately a reflection of the enhanced mobility of financial capital. Yet for all the economic progress made by the emerging world, the developed world, consisting of high-income countries, has strengthened its grip on the world economy. The world's leading financial centres are still New York, London and Tokyo. Emerging nations are, in investment terms, an adjunct to the making of money in the developed world. Indeed, as seen in London, super-wealthy individuals from the emerging world, such as Russians, Saudis and Chinese, are very willing to invest in real estate in the developed world. There is still little solid investment beyond manufacturing industry in the emerging world.

Knowledge check 12

What is meant by the term 'Fairtrade'?

Global trade talks

One characteristic of global trade talks over the last 50 years under the auspices of the WTO has been the degree to which they have stumbled towards agreement. Over the years, the main stumbling blocks have been that:

- the poor countries want a much greater reduction in subsidies for farmers in rich countries so their own farm produce can compete on world markets
- the rich countries want the poor countries to remove import levies on agricultural goods coming into poor countries

Subsidy reductions would have meant some hardships for producers such as US cotton farmers and EU dairy and sugar farmers as they would be threatened by cheaper imports from the developing world. Poor countries in turn worry that if they remove their trade levies, their own farmers would never be able to compete with cheaper imported agricultural products from developed countries. In addition, poorer countries get a large proportion of their tax revenues from taxing imports.

Some countries are very poor and need help to develop their trade. This includes protection of their fledgling processing industries from cheaper imports of processed goods, and richer countries sharing their technical expertise and knowledge to bring them up to twenty-first century trading standards. These countries needed economic support to attain the MDGs, and will need it to attain the SDGs in the future.

The 'global commons'

The term '**the global commons**' was first used in the *World Conservation Strategy*, a report on conservation published in 1991 by the International Union for Conservation of Nature and Natural Resources (IUCN) in collaboration with the United Nations Educational, Scientific and Cultural Organization (UNESCO), the UNEP (United Nations Environment Programme) and the World Wildlife Fund (WWF). It stated:

> A commons is a tract of land or water owned or used jointly by the members of a community. The global commons includes those parts of the Earth's surface beyond national jurisdictions – notably the open ocean and the living resources found there, or held in common, notably the atmosphere. The only landmass that may be regarded as part of the global commons is Antarctica.

The report stated that all people on the planet have a right to the benefits of the global commons. It also stated that, bearing in mind the right of all people to sustainable development, the global commons require protection. The protection of one of the major 'global commons' – the atmosphere – has been explained in the AQA A-level Geography Student Guide 1 – Physical Geography. You may want to revisit those sections examining the human interventions in the carbon cycle that seek to mitigate climate change.

More recently, the internet and the resultant notion of cyberspace have been linked to the concept of the global commons. It will be interesting to see if global governance is, or can be, enforced for this aspect of human living.

Management of the global commons

The key challenge of the concept of the 'global commons' is the design of governance structures and management systems capable of addressing the complexity of multiple public and private interests. Management of the global commons requires a range of legal entities, usually international and supranational, public and private, structured to match the diversity of interests and the type of resource to be managed. They should be stringent but with adequate incentives to ensure compliance. The purpose of such global management systems is to avoid a situation whereby the resources held in common become over-exploited.

In general, many of the global commons (the atmosphere, Antarctica) are non-renewable on human time scales. Thus, resource degradation is more likely to be the result of unintended consequences that are unforeseen, not immediately observable, or not easily understood. For example, the carbon dioxide and methane emissions that drive climate change will continue to do so for at least a millennium after they enter the atmosphere, but species extinctions last forever.

Several environmental protocols have been established as a form of international law. These have tended to be intergovernmental documents intended as legally binding with a primary stated purpose of preventing or managing human impacts on natural resources. However, environmental protocols are not a panacea for global commons issues. Often they are slow to produce the desired effects, and lack monitoring and enforcement.

Global commons The Earth's shared resources such as the deep oceans, the atmosphere, outer space and Antarctica.

Exam tip
Consider ways in which the internet is being, or can be, managed in the world today, but also think about how difficult this is becoming for authorities.

In summary, the designation of a 'global commons' can mean it is:

■ governed by global treaty which, in theory, prevents individual states harming it

■ free for all to use and for the common good

■ subject to debate as to what each of 'treaty', 'free' and 'common good' actually means

Antarctica as a global commons

The geography of Antarctica and the Southern Ocean

Antarctica is the Earth's most southern continent, containing the geographic South Pole. It is almost entirely south of the Antarctic Circle and is surrounded by the Southern Ocean. (Note: the AQA specification includes the Southern Ocean as far north as the Antarctic Convergence). Its size is estimated to be 14 million km², making it the fifth-largest continent. It is twice the size of Australia. A total of 98% of the land area is covered by ice, which averages almost 2 km in thickness, and this ice extends to all but the most northern reaches of the Antarctic Peninsula (AP: see Figure 2).

Knowledge check 13

What is the Antarctic Convergence?

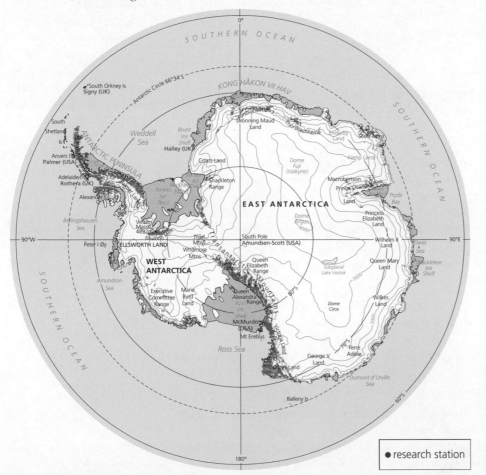

Figure 2 Physical map of Antarctica

Antarctica is the coldest, driest and windiest of all of the Earth's continents, and has the highest average elevation. Climatically, it is a desert, with an annual precipitation of only 200 mm along the coast, with lower totals inland. Around the coasts, temperatures are generally close to freezing in the summer months (December–February), or even slightly positive in the northern part of the AP. During winter, monthly mean temperatures at coastal stations are between –10°C and –30°C but temperatures may briefly rise towards freezing point when winter storms bring warm air towards the Antarctic coast.

Conditions on the high interior plateau are much colder as a result of its higher elevation, higher latitude and greater distance from the ocean. Here, summer temperatures struggle to get above –20°C and monthly means fall below –60°C in winter.

The distribution of precipitation over Antarctica is varied, with several metres of snow falling each year near the coast, but the interior only getting an annual snowfall of a few centimetres. After the snow has fallen it will be redistributed by the winds, particularly in the coastal areas where downslope **katabatic winds** blow. The continent generally experiences moderate winds, with mean wind speeds of around 6 ms^{-1}, but in gales, which can occur on over 40 days a year, mean speeds can exceed 30 ms^{-1} with gusts of over 40 ms^{-1}. The strong katabatic winds, caused by the flow of cold air off the central plateau, make some coastal sites around Antarctica the windiest places in the world.

There are no permanent human residents, but between 1,000 and 5,000 scientists live there at any one time at research stations scattered across the continent. There is a variety of flora and fauna present, consisting of algae, bacteria, fungi, some plants and animals ranging in size from mites and nematodes to penguins and seals. Where vegetation does occur, it can be classed as tundra.

Antarctica is governed by parties to the Antarctic Treaty System (ATS), each of which has consulting status. The Antarctic Treaty was signed in 1959 by 12 countries, and since then a further 38, including India and China, have signed. The treaty prohibits military activities and mineral mining, nuclear explosions and nuclear waste disposal. It supports scientific research, and protects the continent's ecology. The intention of the treaty is to protect the continent's vulnerability to the threats from both economic pressures and environmental change.

Threats to Antarctica

Climate change

According to the Intergovernmental Panel on Climate Change (IPCC), Antarctica is facing the loss of ice from its huge ice sheet, especially from the AP (Figure 2). Furthermore, there is growing evidence of the ice shelves around the continent becoming unstable – for example, large cracks have appeared in the Ross Ice Shelf.

The impact of climate change will vary across the continent (Figure 2):

■ The East Antarctic Ice Sheet (EAIS) is thought to be very stable due to its extremely cold temperatures. If the temperature increased by a few degrees it would still be far too cold for surface melting, and so the ice sheet would not shrink. Only if the temperature went up by huge amounts (tens of degrees) would it

---

Exam tip

Research the climate data for one or more sites on Antarctica, such as Vostok, one of the coldest places on Earth. This will help you to illustrate your answers on the topic.

Katabatic winds Winds that blow down valley sides and valley floors.

Exam tip

Keep up to date with what the IPCC states about climate change in Antarctica to support your answer on the topic.

be possible for major melting to begin. Most scientists working on the Antarctic ice sheets think that the EAIS will not collapse or cause a significant sea-level rise for many centuries to come.

■ The West Antarctic Ice Sheet (WAIS) is the opposite – it is unstable. The fact that much of the ice sits below sea level means that it is sensitive to small rises in sea level, which can cause it to thin. Moreover, the WAIS is drained by several ice streams – fast-moving 'rivers' of ice very different from the slow-moving ice of the rest of the Antarctic ice sheets on the EAIS. Because they move so fast, and drain so much of the ice in the WAIS, the ice streams have the potential to rapidly increase the amount of ice being lost from the ice sheet to the ocean.

■ The AP is one of the most rapidly warming places in the world. Air temperatures here have increased by 3°C over the last 50 years. This warming has been associated with the strengthening of the winds that encircle Antarctica, which in turn drives changes in oceanic circulation and increased upwelling of circumpolar deep water within the Southern Ocean. Across the AP, some 90% of glaciers are receding. The most pronounced impact has been the collapse of some AP ice shelves. Warmth has caused extra melting on the surface of the ice shelves, and eventually this leads to break-up.

Fishing and whaling

Whaling for meat oil and whalebone caused whaling stations for meat processing to be set up in the late nineteenth century on the islands of South Georgia and South Shetland in the Southern Ocean. As whales, especially the blue whale, began to die out many countries stopped the trade.

Fishing became the main economic use of the seas during the 1960s and 1970s, during which large Russian and Japanese fishing fleets fished rock cod and krill stocks using large industrial-scale trawlers. However, by the late 1970s these fisheries had collapsed. Subsequently through the 1980s and to the present day the dominant exploited biomass is krill.

All these fisheries are now regulated through the Convention on the Conservation of Antarctic Marine Living Resources (CCAMLR), which came into existence in 1982. A central objective of Antarctic marine research is to provide scientific support for the work of CCAMLR. It is suggested that the place of krill in the ecology of the area is crucial – if the krill disappear, the whole of the food chain will collapse.

Recently, controversy has arisen between members of the ATS. China and Russia favour greater exploitation of fishing resources, whereas other parties, such as the UN and environmental NGOs, want to see the establishment of marine protected areas (MPAs) which seek to conserve biodiversity. One such MPA was agreed for the Ross Sea in 2017.

The search for mineral resources

The initial Antarctic Treaty did not address the regulation of mineral resource activities. There are many resources as yet undeveloped on Antarctica including oil, coal and many metals. The UK and New Zealand first raised the issue of control of possible future mining within the ATS in 1970, as mineral companies had approached the two countries regarding possible commercial geophysical exploration in the Southern Ocean. The regulation of mineral activities became a very controversial issue.

Exam tip

Use an atlas to locate and assess the relative sizes of the EAIS, WAIS and Antarctic Peninsula, so you can locate them on a map in an exam.

Making connections

The human impact on ecosystems, such as fishing and whaling in the Southern Ocean, is an important element of the Ecosystems option in the Physical Geography course.

Knowledge check 14

What are krill and why are they important for the food chain?

A new Protocol on Environmental Protection to the Antarctic Treaty (known as the Madrid Protocol) was eventually signed in 1991. The Madrid Protocol bans all mineral resource activities in Antarctica (other than for scientific research). The ban can be revisited in 2048, although some countries (e.g. USA, Russia and China) have stated that they want to revisit the ban sooner.

Tourism and scientific research

The development of small-scale tourism began in Antarctica in the 1950s, with commercial tour operators providing passenger ships. The first specially designed, ice-strengthened cruise ship, the Lindblad Explorer, visited in 1969. Since then, the industry has grown considerably, with numbers of tourists increasing from under 9,000 in 1992/93 to over 55,000 in 2018/19. Tourists go to see the wildlife and, as this activity is relatively small scale, it could be argued to be sustainable. However, the summer is the breeding season for most of the wildlife they visit and any disturbance could upset the balance. There is also pressure on the landing sites which receive most tourists, especially the old whaling stations and historic sites such as McMurdo Sound, where the original huts from Scott's expedition in 1912 are located.

All tour operators providing visits to Antarctica are members of the International Association of Antarctica Tour Operators (IAATO), which seeks to ensure that tourism in Antarctica is conducted in an environmentally friendly way. The British Antarctic Survey (BAS) also welcomes a small number of visits to its stations during the austral summer, and groups are given a guided tour of the facilities, where they have the opportunity to learn about the scientific research the BAS undertakes.

Governance of Antarctica

Some aspects of the governance of Antarctica have been mentioned previously, including the ATS and the Madrid Protocol. Further detail is provided here.

Note: as elsewhere in geography, you should also consider how these aspects of global governance underlie and impact your life and the lives of other people across the globe.

The Antarctic Treaty System (ATS) 1959

The ATS is a whole complex of arrangements to ensure

> **in the interests of all mankind that Antarctica shall continue forever to be used exclusively for peaceful purposes and shall not become the scene or object of international discord.**

It prohibits 'any measures of a military nature' but does 'not prevent the use of military personnel or equipment for scientific research or for any other peaceful purpose'.

In 2004, the permanent secretariat to the ATS commenced its work in Buenos Aires, Argentina. The ATS covers the area south of 60°S latitude. Its objectives are simple yet unique in international relations. They are:

- to demilitarise Antarctica, to establish it as a zone free of nuclear tests and the disposal of radioactive waste, and to ensure that it is used for peaceful purposes only
- to promote international scientific cooperation in Antarctica
- to set aside disputes over territorial sovereignty

Exam tip

Research the British Antarctic Survey (BAS), and in particular find out where it has bases on Antarctica, so you can refer to them in your answers.

The Madrid Protocol 1991

See page 33.

The International Whaling Commission (IWC) and Moratorium (IWM)

The IWC:

■ seeks to conserve whale stocks by protection, catch limits and size restrictions
■ designates whale sanctuaries in the Southern Ocean

The IWM:

■ declared a pause in commercial whaling
■ is still in place, although Japan evades it by 'special permit', and Norway and Iceland object to it
■ allows 'aboriginal subsistence' whaling in Greenland and Alaska

The United Nations Environment Programme (UNEP)

The United Nations Environment Programme's (UNEP) direct involvement in Antarctic matters includes the preparation of a regular report for the UN Secretary General on Antarctica. In order to keep the international community informed on the activities of the Antarctic Treaty parties, the UN was requested to serve as a neutral channel to provide information on Antarctic activities. To this end, the UN Secretary General submits to the UN General Assembly a report on the 'Question of Antarctica' on a periodical basis, usually every three years. UNEP prepares the report.

The role of non-governmental organisations

Several non-governmental organisations (NGOs) have an active interest in the protection of the Antarctic and its surrounding ocean and islands. Due to the constraints of the ATS, they can undertake very little in terms of direct impact. Their major involvement, therefore, has been to ensure that the various protocols and regulatory bodies mentioned above are enforced, and that they are active in monitoring threats and enhancing protection.

Knowledge check 15

What is the role of the International Whaling Commission?

Exam tip

Research the work of NGOs such as the Scientific Committee on Antarctic Research (SCAR), the Antarctic and Southern Ocean Coalition (ASOC) and the Antarctic Oceans Alliance, so you can refer to them in your answers.

Summary

After studying this topic, you should be able to:

■ understand the emergence and developing role of various forms of global governance
■ evaluate some issues such as trade and security associated with attempts at global governance
■ know and appreciate the concept of the 'global commons' and illustrate it with particular reference to Antarctica
■ know the contemporary geography of Antarctica and appreciate its vulnerability to a variety of environmental and economic threats
■ analyse the threats faced by Antarctica, such as climate change and tourism
■ evaluate the various forms of global governance of Antarctica, either at a global scale or through NGOs
■ reflect on the geographical consequences of global governance and how they impact your life and the lives of others

■ Changing places

There are two sections to the work to be undertaken for this part of your A-level course:

- a theoretical investigation of place, and
- application of that theory to real places

You are required to examine in detail two such places:

- a local place where you live or study, and
- a contrasting and/or distant place – this place could either be in the same or a different country from the first place, but it must show a significant contrast in terms of economic development and/or population density and/or cultural background and/or systems of political and economic organisation

Some suggest that the better way to tackle this work is to undertake the second part – the investigation into real places – first. This book will examine the theory first, and then the investigations, following the sequence provided by the AQA specification.

For your investigations, the size of a place is defined as a locality, neighbourhood or small community, either urban or rural. You should consider an area you can walk around in 2 hours.

The nature and importance of places

The concept and importance of place

There are many ways to consider the concept of 'place'. 'Place' is where someone was brought up, lives and may eventually die, and contains features that are unique to each individual. A place is more than just a **location**. Try this technique: ask each of your friends, family and acquaintants what your 'place' means to them. We all see places differently.

> **Location** A point in space with specific links to other points in space.

The definition of place, like any concept, is contested. At its heart, though, lies the notion of a meaningful segment of geographical space. We tend to think of places as settlements, for example, Doncaster or Dudley or Dartford. We also consider areas of cities or neighbourhoods – Harrow (London) or Longsight (Manchester) – to be places. Closer in, well-known public spaces are referred to as places – Covent Garden or Land's End. We may refer to a restaurant or café as a 'favourite place'. We also use expressions such as 'knowing one's place' or being 'put in our place' to suggest a more abstract and less locatable interaction of the social and the geographical. We have 'places' set at the dinner table and usually one of them is ours. We may have a favourite chair as our 'place'. We often have our favourite 'places' at school. Places, therefore, create an important basis of life, known as a 'lived experience'.

> **Exam tip**
>
> Create in your notes a series of 'places' that are important, or well-known, at a variety of scales. They will be useful to support your arguments.

The latter notions begin to connect with the importance of place in our lives and experiences. At this point can 'place' take on a larger scale? Is 'place' important to us regionally – some people are proud to be Cornish, or Lancastrians – or nationally – the Welsh and the Scots? What of our 'place' in Europe? How many of us regard ourselves as European rather than British? Beyond the scale of the nation, environmental activist groups work to make us think of the Earth as a place – as a home for humanity – rather than a space to be exploited. Place, then, is not scale-specific. It can be as small as a setting at a table and as large as the Earth. The common assumption that place is a settlement is but one definition of place, and not the most interesting.

So, we must also consider the subjective aspects of a place, and not just the objective. This leads us to refer to '**a sense of place**'. This refers to the feelings evoked by a place for both the **insiders** (people who live there) and **outsiders** (people who visit the area).

Places, then, are particular combinations of material things that occupy a particular segment of space and have sets of meanings attached to them. The concept of place can, therefore, be summarised as:

Place = location + meaning

Insider and outsider perspectives on place

Geographers studying place refer to **insider** and **outsider** perspectives.

Insider perspectives:

- develop through everyday experiences in familiar settings – daily rhythms (e.g. the school run) and shared experiences (e.g. socialising at the village pub)
- are based on experiences acquired over time
- underpin the subjectivity that is the basis for the sense of place of a community
- are intimate/personal views

Outsider perspectives:

- are often about looking and learning – a personal view of entering a location or landscape and discovering that place, as a visitor
- see things afresh
- ask questions that the inhabitants don't think to ask because the answers are so familiar
- are neutral/abstract views

Categories of place

Various cultural geographers have tried to categorise places. Some categories include '**far places**', '**near places**', '**experienced places**' and '**media places**'. Perhaps these are best illustrated by the following.

Consider a village near Banbury in rural 'middle' England where people of different social groups live. At one extreme of the community there are highly educated academics, scientists and business people, whose work is based in the nearby city of Oxford, though they all have computers with high-speed broadband at home. These are in constant contact with, and physically travelling between, colleagues and customers all around the world. The spaces that they move in, both physically and virtually, are thoroughly global (**far places**). At the other extreme of the scale are people who have never been to London and only rarely have made it as far as Oxford,

A sense of place

The personal feelings associated with living in a place.

Knowledge check 16

The term 'locale' is sometimes used in place geography. What does it mean?

Exam tip

You may be asked to define and/or elaborate on each of these terms. Compile a table that summarises their main elements.

in order to go to the shops or maybe to the hospital. Members of this group are known as the 'locals', and most of them work on farms or in village shops and services (**near places**).

Other people in these villages work more or less locally, but are employed as cleaners or caterers by multinational firms, for which this is just one group of workers among many scattered over the globe. Finally, there are the partners of the academics, scientists and business people, several of whom are occupied in a daily round of nurseries and child-minders, often being the heart and soul of local meetings and charities (**near places**). They tend to drive into Oxford to do their shopping, maintain contacts with extended family outside the local area and like to go on holiday to somewhere 'exotic' (**far places**).

This account of the different social groups in this hypothetical village shows how place is far more permeable for some (the wealthy incomers and out-goers) than it is for others. However, even the more rooted, less travelled lower income people here are increasingly touched by wider events. Farm workers, for example, are subject to agricultural policy decisions made in London or elsewhere, and the cleaners and caterers who work for multinational firms in the area might well feel the force of global economics if those companies were to cut back on jobs (all **experienced places**).

Add to these the assertion that in today's electronic society people have no 'sense of place'; rather they occupy **media places**. Electronic media are undermining the traditional relationship between a physical setting and a social setting. The world's media bring to our location events that are taking place in another location, and hence in some ways we are transported to that location even though we actually remain in our location. At a much smaller scale is the situation in which two people are having a telephone, Skype or Zoom conversation in two different locations. Indeed, the telephone (or computer) brings them closer together than with other people in their respective locations. Furthermore, the plethora of media communications in recent years has forced us to consider the bias or agenda of the source(s).

Factors contributing to the character of places

Endogenous factors include:

- the natural characteristics of a place – geology, altitude, topography, coastal or inland
- demographic features – number of inhabitants, their ages, gender and ethnicity
- socioeconomic characteristics – types of families (young adults, families with children, retired people), income levels, level of education, types of employment (e.g. skilled, manual, professional)
- cultural factors – religious groups, local traditions
- political factors – type of local, regional or national government; local groups such as residents' associations and campaigning groups
- the built environment – the ages/styles of buildings and their building materials; flats, terraced, semi-detached or detached housing; building density

Exam tip

Any location can satisfy one or more of these categories of place, as they all depend on individual viewpoints and situations. Be aware of the complexity of this topic.

Knowledge check 17

All of these categories of place owe their origin to the ideas of Yi-Fu Tuan. He developed the phrase 'field of care'. What is the 'field of care'?

Endogenous factors
Factors that are caused or originate from within, i.e. internally.

Exogenous factors include:

■ features associated with globalisation – clothing shops, fast-food outlets, places of employment, social media

■ evidence of multi-ethnicity caused by inward migration

Your studies of 'place'

At the heart of this unit of work are the two detailed case studies:

■ one local place where you live or study

■ one further contrasting place

When reading the following two sections (Relationships and connections and Meaning and representation), ensure that you appreciate how the factors mentioned affect continuity and change in your chosen places. Also, reflect on how your life and that of others has been and is affected by such continuity and change in your places.

Relationships and connections

Shifting flows

People

Shifting flows of people help shape a place profile, due to:

■ **migration**: people move into a place or they move away; these flows of people can change the demography of a place

■ **life-cycle stage**:
  – an area can receive an influx of young adults, e.g. students or young professionals, living away from home
  – an area can be affected by an influx of commuters due to accessible railway lines and stations
  – an area can receive an influx of retired people, e.g. some UK coastal towns

Resources

Resources can shape a place profile.

■ The availability of a mineral resource, e.g. coal or iron ore, can lead to the establishment of a mining community.

■ If that resource runs out or is no longer required, the place undergoes change, although evidence of its previous profile persists.

■ The rise in electronic communication has given great significance to the connectivity a place has via the internet and mobile technology, e.g. remote rural locations.

Money and investment

Money and investment help shape a place profile, with:

■ funds provided by governments, TNCs and local authorities (LAs)

■ spending on characteristics of places such as transport infrastructure, education, health and environment

Exogenous factors
Factors that are caused or have an origin from without, i.e. externally.

Making connections

The importance and role of endogenous and exogenous factors can be linked to the impact, and management, of hazardous events and/or environmental changes in a place.

Exam tip

Consider the place where you live. What are the endogenous and exogenous factors that have contributed to its growth? You can use these to illustrate exam answers.

Migration A permanent change in where a person lives – can be internal or international.

Life-cycle stage Describes the age and family status of a person such as young adult, or a person who is married with children or retired.

Resource A broad definition is needed here – anything that people can make use of.

- small-scale funds from individuals or groups of individuals that can result in the creation or modification of businesses, e.g. attempts to resist **clone town high streets**

Clone town high street One dominated by chain shops which can be found all over a country; few local, independent businesses survive.

External forces

A number of large-scale **agents of change** can affect a place profile – governments, TNCs and international/global institutions. They often act together to change the character of a place, and so it is difficult to separate out their relative impacts.

- **Governments:** invest in the infrastructure of a place, thereby enabling change; LAs give enticements to others to invest in the locality.
- **TNCs:** provide investment that gives them the greatest return, e.g. a factory, or hotel complex.
- **The EU:** has provided vast sums of money to regenerate declining areas across the UK.

Past and present connections

Places should also be studied in their context, for example how past and present connections have shaped them and embedded them in regional, national, international and global contexts. The impact of this will vary from place to place; all places have a history that has shaped their development, but that history may be more significant for some places than others. Another key point here is to appreciate the complexity of the human environment. As geographers, we are interested in more than just the economic significance of places. They are places of political, cultural and social significance too. Politics, religion, culture and history all influence the make-up of a place. In fact, geographers often describe places as being made up of a series of layers, or as a **palimpsest**, which can be unravelled in order to develop a greater appreciation of the place. For example, Istanbul is best understood when we think about the layers of history that influence the cityscape, from its links with the Greeks and the Ottoman Empire to its contemporary relationships with Europe and Asia.

At a smaller scale, the state, and even local government, can use place, as space can be used to assert power and ideology. Many capital cities and smaller cities and towns have public places, such as large open streets, parks and squares, which have played a role in asserting power and maintaining order. At varying times, these public places have also had a non-political role that also defines the place. Past connections can also be evidenced by the ethnic composition of the population, old buildings, street names and even the place's name.

Meaning and representation

Attachments to place

This section examines how people perceive, engage with and form attachments to places. It also looks at how places present themselves to others and how they

Knowledge check 18

To what extent does globalisation impact on place?

Agents of change Individuals, groups, multinational corporations, institutions (national or international), media and governments that have driven change either intentionally or unintentionally.

Making connections

The impact of several topics studied under Global systems will be relevant here, for example TNCs and economic/trade groupings.

Palimpsest Something that has been reused or altered but still bears visible traces of its earlier form.

Exam tip

Consider the place where you live. What are the public places created in your area, and what was/ is their purpose? You can use these to illustrate exam answers.

Knowledge check 19

Having completed this part of the topic, now distinguish between 'location' and 'a sense of place'.

are represented to the rest of the world. Key concepts within this area of study are **identity**, **belonging**, **ownership** and **wellbeing** in relation to place. Perspectives and experiences of place are also key features of this area of study.

Identity and belonging are often judged by external and internal characteristics of the person, such as age, gender, race, sexuality, marital status, religion and physical (dis)ability:

Age
- Perceptions of a place often change through our life cycle.
- Young adults may prefer to live in locations where work, shops and leisure facilities are close by.
- People with young families may desire more space (a garden, a park), with access to a nursery or school.
- Older people may prefer a more secluded place.

Gender
- Perhaps not as influential in developed societies as in the past, though stereotypical attitudes may prevail regarding work vs home.
- Some workplaces are still dominated by one gender, e.g. mining, nursing.
- The perception of 'safety' (or lack of it) in some urban areas still impacts on gender, e.g. train carriages at night.

Ethnicity and sexuality
- Places with concentrations of particular ethnicities have developed for many years.
- The character of the place may have changed to reflect the place of origin of the ethnic group, e.g. 'Chinatowns' in cities.
- Shops and services (e.g. religious) are also influenced by these groups.
- Some places in cities have concentrations of LGBTQ+ groups.
- For all of the above, the sense of belonging is enhanced, and provides security.

Some aspects of identity are political and emotional, and bring about feelings of being part of a group – belonging or otherwise. Inclusion or exclusion can then lead to feelings of discrimination, prejudice or injustice. Issues such as racism, sexism, homophobia and elitism are associated with identity and belonging. These are often both reflected and reinforced in space and place. Think, for example, of the Canal Street Gay Village of Manchester, or the 'posh' district of Mayfair in London.

Identity therefore is formed in relation to others – it is said to be socially constructed. We become aware of who we are through a sense of shared identity with others (such as speaking the same language, or having the same political views), and by a process of setting ourselves apart from those we consider different from ourselves. Identity can also have a degree of fluidity, or flux. With time, our identities may change as, for example, our social class changes.

Ownership refers to the feeling of being in possession of a set of values, or a particular identity and, if desirable, it then adds to our sense of wellbeing and worth. Once again, this takes us into the role of processes of inclusion and exclusion. The connection between place and particular meanings and identities leads to the notion of places where it is possible to be either 'in place' or 'out of place' – an insider or an outsider.

Identity
An assemblage of personal characteristics such as gender, sexuality, race and religion.

Belonging A sense of being part of a collective identity.

Ownership The feeling of being in possession of a set of values, or a particular identity.

Wellbeing The positive outcome of a shared identity, and a sense of belonging.

Exam tip

Consider a town/city you know well. Think about the location of places where different groups of people might live or meet, and their reasons for doing so.

Exam tip

Perception is an important influence on the meanings people give to places. Make sure you have real-world examples to support your knowledge and understanding.

Knowledge check 20

Identify some of the varying identities of people involved in the European migration crisis of the last decade.

Things, practices and people labelled as out of place are said to have transgressed often invisible boundaries that define what is appropriate and what is inappropriate. Young people gather on street corners or skateboard on street furniture; the homeless find ways to live in inhospitable places; street artists (such as the well-known Banksy) redecorate urban walls to establish new meanings and identities in those communities. On a much larger scale, the mass migrations of refugees that took place across Europe in the latter part of the last decade challenged the whole concept of boundaries, and national place.

In summary, attachments to place:

- can come with long-term association with a place – based on memories
- can exist even though a person no longer lives there
- are more subjective – a sense of 'homeland' with a strong feeling of identity, but with no geographical or political basis

Influencing and creating place-meanings

Some organisations attempt to manage, or even manipulate, the perception of place for their own ends. This is not always as sinister as it may seem, but instead is aimed at managing how others see work being undertaken in a community in order to improve the place, such as to create a more homogeneous identity, or to raise awareness of what is being done. External agencies such as community groups dealing with local health issues, corporate bodies who are keen to raise awareness of new developments taking place, and local and national governments all try to manipulate place perception to varying degrees to achieve further policy ends.

Perhaps the best example of all three of these types of bodies working together is in a **partnership** of non-governmental and governmental organisations aimed at developing or regenerating an area. Examples include Cardiff Bay (Europe's largest waterfront development, in the Welsh capital), Salford Quays (another waterfront development, in Greater Manchester) and the Ocean Gateway Project (sometimes referred to as the Atlantic Gateway, which will involve extensive redevelopment of the Port of Liverpool and the Manchester Ship Canal).

The attraction of partnerships results from their apparent potential to bring interested local organisations, including business, and agents of government together to pool their resources (financial, practical, material or symbolic), leading to the development of joint and consensual strategies to address issues in that place. They blend together the public, private and voluntary sectors. It is important, therefore, that they manipulate the perceptions of residents of the area to their common sense of good for the area.

To this end therefore, any form of external agency (government, corporate body, community or local group) can seek to influence or create specific place-meanings, and thereby shape the actions and behaviours of individuals, groups, businesses and institutions. This is sometimes referred to as 're-imaging' and/or 'rebranding'.

Consider the example of Liverpool. Culture (popular music, the arts, sport) has dominated its recent rebranding. Liverpool has a rich history in popular music (notably The Beatles), and the performing and visual arts. It also has two Premier League football teams. Since 2003, when Liverpool was awarded the status of European Capital of Culture for 2008, the city centre has been transformed through

Exam tip

You may be asked to define and/or elaborate on the terms 'identity' and 'representation'. Compile a table that summarises their main elements.

Partnership

Non-governmental and governmental organisations working together to develop or regenerate an area and give it a new identity.

Exam tip

Research one or more of these types of example and examine how the redevelopment partnerships manage the perception of place.

major investment. In addition to the nearly £4 billion invested in regeneration, the city's economy is said to have been boosted by an annual £1 billion of additional income. Over 15 million visitors were attracted to the city for the 7,000 cultural events it hosted in 2008. Attendance at the venues within the Albert Dock increased by over 30% and there were record visitor numbers across all of the city's attractions.

In some ways, Liverpool's rebranding has made the city centre more similar to other city centres but the city is still able to promote its distinctive cultural and maritime character. In recent years, up to 10 million tourists have visited Liverpool each year from the UK, other European countries and further afield, especially China, Japan and the USA. The budget airlines that use Liverpool's John Lennon Airport have boosted tourism to the city. In fact, these tourists have made Liverpool one of the ten most visited destinations in the UK. A variety of agencies have managed much of this increased popularity.

Rebranding may, however, have negative impacts. The regeneration of Cardiff Bay, from the formerly named Tiger Bay, has been argued by some to have diluted the part played in its original growth by ethnic groups who had settled in the area.

In summary, creating a place-meaning can involve:
- community groups seeking to create a 'village' spirit
- local authorities seeking to market their place to create a positive perception for investment
- corporate bodies who are keen to raise awareness of new developments taking place
- national governments manipulating place perception to achieve further policy aims, often in partnership

Representation of place

Places may be represented in a variety of different forms such as advertising copy, tourist agency material or local art exhibitions and in diverse media (e.g. film, photography, art, story, song). These may often give contrasting images to that presented formally or statistically such as in cartography and census data.

Artistic methods

Key features:
- Several agencies make use of our imagination to influence how we see a place.
- Advertising agencies combine visual and written imagery to enhance the settings of a place.
- Tourist boards may select aspects of a place that may fit a desired perception of that place. This concept is not new – in the early twentieth century, railway companies produced many posters to advertise faraway places, accessible by train.
- Some **representation of place** may be informal, but no less powerful – novels, poems, songs, the visual arts and other diverse media (television, film, video, photography) all 'bring alive' different places.
- The impact of television on the tourist prospects of several locations can be highlighted here – think of the impact of programmes such as *Hollyoaks*, *Made in Chelsea* and *Game of Thrones* on their respective locations of Chester, West London and Northern Ireland/Iceland.
- In these cases, we may struggle to match the fictional with the factual.

Knowledge check 21

Distinguish between rebranding, re-imaging and regeneration.

Representation of place The cultural practices by which human societies interpret and portray the world around them and present themselves to others.

Exam tip

Consider how twenty-first-century advertisers entice consumers to tourist locations, to add depth to your answers.

Exam tip

All of these locations are UK-based. Think of several overseas locations in which various forms of media have highlighted their worth (or otherwise).

Formal methods

Key features:

■ More data about places are now collected, stored and analysed than ever before.

■ In many countries, the most effective formal representation of places is their census.

■ There has been a dramatic increase in the quantity and quality of **geospatial data**.

■ Many government agencies maintain websites that present formal representations of places.

■ Formal representations offer rational perspectives of a place profile, such as numbers of people living in a place, their ages, genders and educational qualifications.

■ They are limited, however, in their ability to indicate 'lived experience' aspects of a place profile.

Past and present processes

Past and present processes of development can be seen to influence the social and economic characteristics of places and hence be implicit in their place-meanings today. See also 'Past and present connections' on page 39.

Some further points:

■ There is an uneven spread of resources, wealth and opportunities within and between places.

■ **Social inequality** tends to be linked to differences in access to housing, healthcare, education and employment.

■ In the UK, these are combined to give an Index of Multiple Deprivation (IMD) (see page 45).

■ Several factors determine social inequality; some have operated in the past, and some in the present:

  – **Levels of income:** in general terms, the higher the level of income, the higher the standard of living; unemployment increases social inequality; both are linked to past industrialisation or deindustrialisation.

  – **Housing:** affordability of housing is crucial, either for rent or purchase; in some rural areas demand created by second-home ownership has caused problems for young adult residents.

  – **Health:** ill-health is associated with sub-standard housing, poor diet and unhealthy lifestyles; access to healthcare is crucial.

  – **Education:** access to education, both primary and secondary, is an issue for the developing world; in the UK there are perceived differences between public and private, inner city and suburban schools.

Place studies

This part of the specification requires you to understand two places through the collection, analysis and interpretation of quantitative and qualitative data, including their representation in the media.

Geospatial data
Locational information about a place – geographic information systems (GIS) are often used.

Social inequality The uneven distribution of opportunities and rewards for different social groups, defined by factors such as age, gender, class, sexuality, religion or ethnicity.

Exam tip

Consider the advantages and disadvantages of methods of measuring social inequality, such as income, education and health.

You must undertake two place studies:

1 Explore the developing character of a place local to your home or place of study.

2 Explore the developing character of a contrasting and/or distant place. This place could be in the same country or a different country but it must show significant contrast in terms of economic development and/or population density and/or cultural background and/or systems of political and economic organisation.

Both place studies must focus on **people's lived experience of the place in the past and at present**, and *either* **changing demographic and cultural characteristics** *or* **economic change and social inequalities**.

'Local place' tips

Choose:

- a place with access to a good supply of data from a range of sources (see below)
- an urban place that has undergone or is undergoing change – this could be socioeconomic (economic decline or regeneration, new housing estate, near an industrial estate or out-of-town shopping centre) or demographic/cultural (inward or outward migration)
- a rural place that has experienced counter-urbanisation, depopulation or a contested landscape (new housing estate, second homes, impacted by infrastructure projects such as HS2)

'Contrasting/distant place' tips

Choose a place:

- that has good availability of data from a range of sources (see below)
- where charities and non-governmental organisations are useful sources of information
- with which there is a possible twinning arrangement with a school, or one where you can make use of video-calling to talk to people
- you have visited, or will visit

These place studies must apply the knowledge acquired through engagement with the specification content and demonstrate understanding of the ways in which your own life and those of others are affected by continuity and change in the nature of places.

Both place studies must use a variety of sources to acquire knowledge and understanding of the places and their changing characters. They must give particular weight to qualitative approaches involved in representing place, and to analysing critically the impacts of different media on place-meanings and perceptions. You must also use quantitative data, including geospatial data, to present place characteristics. Their use should allow the development of critical perspectives on the data categories and approaches.

Suitable data sources could include:

- statistics, such as census data
- maps
- geo-located data

> **Exam tip**
>
> For your investigations, aim for a locality, neighbourhood or small community, either urban or rural. You should consider an area you can walk around in 2 hours.

> **Exam tip**
>
> The specification states that you should focus on *either* changing demographic and cultural characteristics *or* economic change and social inequalities. However, these overlap. It is advised that you **do not** categorise in this way. Study all elements and use the better set of information for each question as it arises.

- geospatial data, including geographic information systems (GIS) applications
- photographs
- text, from varied media
- audio-visual media
- artistic representations
- oral sources, such as interviews, reminiscences and songs

The Index of Multiple Deprivation (IMD)

One highly useful online source of data for places in the UK is the IMD (https://parallel.co.uk/imd). The IMD is published by the Department for Communities and Local Government and informs national and local government decision-making. It ranks the basic areas for the UK census (Super Output Areas, or SOAs) across England according to a combination of seven domains of deprivation: income, employment, education, health, crime, barriers to housing and services, and living environment. Each of these domains is based on a further number of indicators – 37 in total. Each indicator is based on the most recent data available although, in practice, most indicators in the 2019 data, for example, relate to the tax year ending April 2017. It also uses smaller parts of the SOAs, called Lower SOAs (LSOAs), each containing about 1,500 residents or about 650 households, to identify small pockets of deprivation.

The deciles shown on the IMD maps are produced by ranking 32,844 LSOAs and dividing them into ten equal-sized groups. Decile 1 represents the most deprived 10% of areas nationally, and decile 10 the least deprived 10%. When interpreting these data, note:

- The rank of the deciles is relative: they simply show that one area is more deprived than another but not by how much.
- You will see large areas of colour, often the same or similar shade. Note that these show areas and not numbers of people living there.
- The data shown by such neighbourhood-level maps provide a description of an area as a whole and not of individuals within that area.
- Deprivation, not affluence, is mapped.

Exam tip

You could examine alternative perspectives on place characteristics, such as the principles of 'Doughnut Economics' as featured in the work of Kate Raworth. Here sustainability and living space are emphasised.

How do you assess the character of a place?

An obvious method of exploring the character of a place is by direct observation, especially in the place local to you. Walk through the area, look and simultaneously question what you see. Try to relate, interpret and assess your observations, i.e. form an opinion about 'what is good' and 'what is wrong', and map them.

While reading the environment you walk through, identify aspects you consider important to be changed (problems or issues) and those which are very characteristic of the place and should be maintained. It is important that you engage with the place on a personal level.

You should also research further endogenous and exogenous information about the place, not acquired by your personal reading of its character, such as:

- physical morphology of the landscape
- history of the evolution of the place

- socioeconomic characteristics of the population of the area (possibly employing census data), including any people who may have come from overseas
- evidence and dynamics of change in the area
- aspects of connectivity to the area, such as any transport systems
- how the area is featured in media, such as local newspapers

Your objective should be to aim to get a comprehensive, interpretive and critical portrait of the place – a quick and meaningful sketch of the present situation and its dynamics towards change. Information can be mapped, labelled and illustrated (through photographs, sketches or verbal/video records) in order to describe, clearly and spatially, the results of your observations. You could use geospatial mechanisms and other forms of GIS (such as ESRI/ArcGIS).

Two or more people could also undertake direct observation of the character of the area. Multiplying the number of observers enriches the quality of the observation as long as they complete the task independently. By comparing and discussing results, you control, to some extent, deviation resulting from any individual subjectivity and observational weaknesses. You could ensure that you obtain updated maps and aerial photographs of the area (online platforms being very useful here), and start to collect cartography of the area produced over time, again possibly from online sources. This will show how the place has developed over time, and created its palimpsest.

Some themes to develop in the study of the historical evolution of the place could include:
- identification of the sequence of formation, filling and growth of the place
- development of its road layout
- details of significant events and agents in structuring the processes of this evolution
- evolution of the social fabric and ownership of the place, by its population and their activities
- the ages and types of buildings, and their state of preservation
- current building types, and ways in which they may be grouped or aggregated (e.g. housing types) by use or design
- historical and artistic ownership and/or representation, for example, in paintings and old photographs

Regarding the transport facilities in the place, you could identify:
- the main directions for public and private transport and any terminals/stops
- points of higher traffic intensity and possible conflict relating to the movement of cars, and potentially with pedestrians or cyclists

Conducting interviews with people who know the area well (such as people connected with parish churches, health services, police officers and long-time inhabitants) is also important. All of these research methods will help you to formulate a mental map of the area.

Another interesting aspect that could be investigated concerns the presence of plant and animal species, through assessment of the proportion of the 'green' in relation to the 'built up'. Vegetation surveys can take several different forms and involve varying functions, for example, gardens or public spaces, and can form part of the physical character that the place offers. You could consider the amount of open public space, and the degree to which landscaping has taken place.

Essentially, you are getting the place's 'portrait' at the time of the observation.

Summary

After studying this topic, you should be able to:

■ understand the nature and importance of place in human life and experiences
■ appreciate the various perspectives and experiences of place
■ explain how places develop meaning for their inhabitants, and evaluate the various ways in which places can be represented
■ appreciate that a number of external agencies seek to manage and manipulate perceptions of place, mostly for positive reasons
■ understand the factors that contribute to the developing character of places, some of which are endogenous and others exogenous
■ analyse the forces that operate to change a place, and yet appreciate that all places have a context
■ carry out two detailed studies of places at a local scale, as required by the specification

■ Contemporary urban environments

Urbanisation

The world's population is increasingly concentrated in urban settlements, presenting both opportunities for and challenges to sustainable development. Cities drive economic and social development as hubs of commerce, transportation, communication and government. But rapid, unplanned urban growth can lead to an expansion of urban slums, exacerbating poverty and inequality, hampering efforts to expand or improve basic infrastructure and deliver essential services, and threatening the environment. By anticipating urban growth, countries can plan for future change and ensure that **urbanisation** remains a positive force for sustainable development.

Urbanisation
The process where the proportion of people living in towns and cities increases.

Global patterns of urbanisation since 1945

The world has urbanised rapidly since 1950 and projections indicate that it will continue to urbanise in the coming decades. In 1950, the world was mostly rural: more than two-thirds of people lived in rural settlements and less than one-third in urban settlements. In 2019, 55% (more than 4 billion) of the global population was urban. This distribution is expected to shift further towards urban areas over the next 35 years so that, by 2050, the world's population will be one-third rural and two-thirds urban, roughly the reverse of the situation in the mid-twentieth century.

Exam tip

An obvious point perhaps, but this optional unit requires you to know where the major urban areas of the world are located.

Global urbanisation has been driven by rapid natural growth of the urban population, concurrent with stagnating growth of the rural population, together with inward migration from rural areas. The global urban population has increased by a factor of five, from 0.7 billion in 1950 to 4 billion in 2014. It is expected to increase by another 60% by 2050, when 6.3 billion people are projected to live in urban settlements. The global rural population is ceasing to grow. It is projected to reach a peak of just under 3.4 billion shortly after 2020 and to decline thereafter to 3.2 billion in 2050.

A growing number of countries are becoming highly urbanised, with a majority of their populations concentrated in urban settlements. Only some countries in sub-Saharan Africa (e.g. South Sudan and Kenya) and south and South East Asia (e.g. India and Vietnam) have larger rural populations than urban ones.

By 2050, nearly 70% of countries or areas in the world are projected to be more than 60% urban and almost 40% will be at least 80% urban. The number of countries that are predominantly rural will decline such that by 2050, just 27 countries are projected to be less than 40% urban, half of which are small islands or territories with under 2 million inhabitants.

The levels and pace of urbanisation vary widely across regions and countries. Northern America and Latin America and the Caribbean are the most urbanised regions, with 80% or more of their populations living in urban settlements in 2019. Europe, with 75% of its population living in urban areas in 2019, is expected to be more than 80% urban by 2050. Large parts of Africa and Asia remain mostly rural. However, both regions are projected to urbanise faster than other regions over the coming decades, reaching 56% and 64% urban by 2050. Nevertheless, Africa and Asia are expected to remain the two least urbanised regions of the world.

The scale and speed of urbanisation across the developing world today are unprecedented – creating a number of well-known **mega-cities**, such as Jakarta, Istanbul, São Paulo and Cairo. Others, though, many people will not have heard of. For example, Chengdu in China has a population (including its rural hinterland) of more than 14 million. Similarly, few will have heard much of cities like Ghaziabad, Surat or Faridabad in India, Toluca in Mexico, Palembang in Indonesia or Chittagong, the Bangladeshi port. Each of these cities is among the fastest-growing settlements in the world.

Issues

Urbanisation has resulted in a number of issues:

- The development of **primate cities** in some countries and the growth of marked differences between the economic **core** and **peripheral** regions of a country.
- A city may not be able to provide enough housing for all of the migrants so they are faced with three alternatives: sleep on the streets; try to rent single rooms or houses; or build their own shelters on land which they do not own.
- This has led to the growth of **squatter settlements** or **shanty towns** which have problems of overcrowding, sewage disposal, disease and crime.
- Pressure on services such as refuse collection, health provision, education, police and fire services, power supplies and sewage disposal.
- Transport systems become overused and the road network is unable to accommodate the increase in vehicular traffic.
- The number of jobs available in the city does not match the incoming migration which leads to vast unemployment. In addition, a large number of people work in the **informal sector** and are therefore classified as **underemployed**.

Studies of cities in the developing world, however, have shown that the picture of a bleak downward spiral is not always the case. To many migrants, urban life may be very superior to conditions in rural areas. Employment can be found with better wages and many shanty towns are not always areas of deprivation, extreme poverty and disease. Some are very well-organised and not the first destination of recent immigrants to the city.

Types of urbanisation

A discussion of a variety of forms of urbanisation now follows. You should be aware of the wide range of factors that influence each of these forms. These include economic, social, technological, political and demographic factors, each of which plays a role, but in varying degrees.

Suburbanisation

Suburbanisation has resulted in the outward growth of urban development that has engulfed surrounding villages and rural areas. In the UK, during the mid- to late-twentieth century, this was facilitated by the growth of public transport systems and the increased use of the private car. The presence of railway lines and arterial roads enabled relatively wealthy commuters to live some distance away from their places of

Mega-city A city with more than 10 million people (including their immediate urban areas).

Knowledge check 22

Examine the variation in the pattern of urbanisation in one continent of your choice.

Primate city Where one city in a country dominates the city size distribution – it is more than twice the size of the second-largest.

Shanty town An illegal settlement within a city that contains cheap, often hand-built, houses.

Informal sector Unregulated and unstructured employment.

Exam tip

Questions often ask you to evaluate or assess issues – remember issues are not always problems, there may be benefits too.

work but in the same urban area. Extensive areas of housing were built on the edges of major cities – for example in London, huge suburban estates were built in places such as Wimbledon, Twickenham and Wembley.

The edge of town, where there is more land available for car parking and expansion, also became the favoured location for new offices, factories and shopping outlets. In a number of cases, the 'strict control' of the green belts was ignored (or at best modified) in the light of changing circumstances. More recently, there has been the development of new housing areas on 'previously developed land' (also known as 'brownfield sites'), leading to the infilling of abandoned industrial areas, private gardens and other open spaces in urban areas, such as school playing fields.

Counter-urbanisation

Counter-urbanisation refers to the migration of people from major urban areas to smaller urban settlements and rural areas. There is a clear break between the areas of new growth and the urban area from which the people have moved. As a result, counter-urbanisation *does not* lead to suburban growth, but to growth in rural areas beyond the main city. A number of factors have caused the growth of counter-urbanisation:

- A negative reaction to city life – many people want to escape from the air pollution, dirt and crime of the urban environment.
- Greater affluence and car ownership allow people to commute to work from such areas. Indeed, many sources of employment have also moved out of cities.
- Improvements in technology, such as access to broadband, have allowed more freedom of location and facilitated people working from home and hence not needing to go to the city workplace on every day of the week.
- There has been a rising demand for second homes and earlier retirement into rural areas. The former is a direct consequence of rising levels of affluence.
- The need for rural areas to attract income. Agriculture is facing economic difficulties and one way for farmers to raise money is to sell unwanted land and buildings for development of new houses and building conversions.

Urban resurgence

This refers to the movement of people *back* into urban areas, particularly the inner city or even the **CBD** itself. It is often associated with urban regeneration schemes, urban rebranding schemes or gentrification (Table 6). Increasingly, it is also associated with the move towards sustainable urban communities.

Exam tip

Suburbanisation and counter-urbanisation are often confused – make sure you know the difference between them.

CBD Central business district

Table 6 Forms of urban resurgence

Regeneration	Rebranding	Gentrification
The investment of capital and ideas into a rundown city area to revitalise and renew its economic, social and/or environmental condition.	The process of regenerating a city's economy and physical fabric as well as projecting a new, positive urban image to the wider world.	Housing improvement by people rather than by organisations – associated with a change in neighbourhood composition in which low-income groups are displaced by more affluent people, usually in professional or managerial occupations.

The general effects of urban resurgence include:

- the reuse of old buildings for a new purpose – housing, offices, hotels
- a positive multiplier effect with more investment being attracted
- vibrant city centre activities – the '24-hour' city
- the displacement of lower-income people, often with levels of inequality increasing

Processes involved in urban change

Deindustrialisation

This occurs:

- when there is an absolute decline in the importance of manufacturing in the economy of a country – there is a fall in its contribution to GDP
- where there has been an over-reliance on traditional heavy industries, such as iron and steel, chemicals, shipbuilding and textiles
- due to strong competition from overseas areas where new technology and less unionised practices have been adopted

Decentralisation

This occurs:

- when there is an outward movement of population and industry from established central urban areas towards the suburbs or to smaller urban centres (counter-urbanisation)
- where there is encouragement by government agencies trying to spread investment and development from the central area of a city towards the periphery
- due to the negative aspects of higher crime, noise, pollution and land costs found in central locations

The rise of the service economy

This occurs:

- when deindustrialisation and decentralisation have both encouraged the rise of the service-based activities in developed world cities, such that it becomes the main employer of people in a country
- where there has been an increase in employment in education, health, public transport, retailing, local government, banking and finance

Urban policy and regeneration in Britain since 1979

Since the end of the Second World War, regeneration schemes have taken a variety of forms. Key elements of these schemes were slum clearance and housing renewal, new industrial growth and development, improvements to transport systems and environmental improvements. Similarly, a central theme for many of the later schemes was to encourage private sector investment, instead of that from local councils or central government sources (Table 7). This is because regeneration has become increasingly vulnerable to recession, financial crisis and shortages of credit.

Making connections

Regeneration, rebranding and gentrification are important processes in the study of Changing places.

Knowledge check 23

Give one economic, one social and one environmental impact of deindustrialisation.

Table 7 Summary of urban policy and regeneration in Britain since 1979

Policy	Date	Main features
Pump-priming	1979–1998	Autonomous Urban Development Corporations (UDCs) and Enterprise Zones were given responsibility for infrastructure renewal and derelict site remediation in order to attract private sector investment. UDCs (e.g. the London Docklands) were free from many planning regulations; local and national government involvement was minimal.
Public–Private Partnerships (PPPs)	2000–2010	Regional Development Agencies (RDAs) worked to combine private and public investment to regenerate key sites in cities. Flagship and landmark buildings were used to 'kick-start' investment (e.g. Salford Quays).
Local Enterprise Partnerships (LEPs)	2010–present	The Coalition government abolished most RDAs in 2010. Since then, regeneration has been largely led by private investment often led by retail (e.g. Westfield schemes such as Stratford, London) and private housing developments. The scale of regeneration is much more local. Local councils merely act as advisors and facilitators.

Rebranding is an important element of many regeneration schemes. As with companies that compete with each other to obtain a growing share of the market, cities vie with one another to attract investment and visitors. They also compete to keep their existing residents or attract new ones. Key elements of a city's 'brand' are the urban environment (a city's artefacts such as the cityscape and its buildings), its essence (people's experiences of the city), and its brandscape (how the city positions itself in relation to other cities).

Rebranding often involves the reworking of a city's existing identity(ies) but sometimes includes creating new ones. Every city has many identities and it can be difficult to decide which to promote. A city can represent different things to different people. In some cases, rebranding can counter negative images and encourage the people who live and work in a city, and those who may wish to visit or invest in it, to think about it in a different way. City rebranding is a global phenomenon and has been used by cities such as Liverpool and Barcelona to change their images.

The emergence of mega-cities

Urbanisation can be tracked by the growth of mega-cities (Figure 3). In the 1960s, the world's largest cities were in developed countries. In the 1970s, several Latin American mega-cities emerged. From 1970 to 2000 a few developed world mega-cities, such as London, actually saw population decline due to deindustrialisation and counter-urbanisation. Since 2000, mega-city growth has been centred in Asia, especially India and China. Africa has relatively few mega-cities although many cities here are growing rapidly. In 2015, there were over 30 mega-cities – the number is expected to increase to 40 by 2030. Table 8 gives characteristics of the different types of mega-city.

Making connections

Rebranding is an important process in the study of Changing places.

Exam tip

Urban regeneration is a very topical issue. Keep up to date with any urban regeneration schemes you study, or that are local to you.

Knowledge check 24

Identify two ways in which the UK government has assisted urban regeneration by providing better transport links.

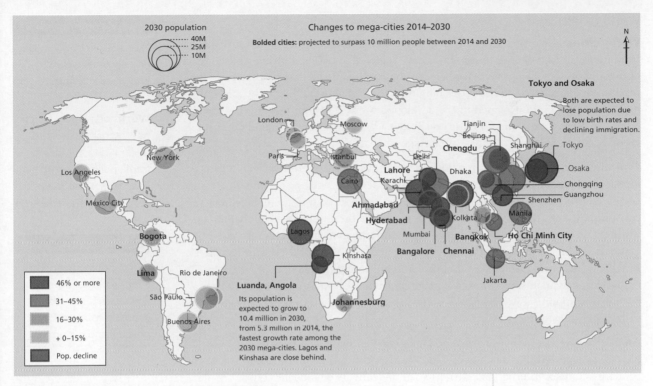

Figure 3 Changes to mega-cities (2014–2030)

Table 8 Characteristics of types of mega-city

Immature	Consolidated	Mature	Established
Grow rapidly in an uncontrolled way; many are in Africa. Growth is fed by rural–urban migration and is so rapid that housing, transport, education, sewers and water services cannot be built to keep pace with growth. This leads to major health, housing and pollution problems. An example is **Lagos** in Nigeria, which has an annual population growth rate of 4.5%, 70% of the population living in slums, and only 20% of houses having direct access to a piped water supply.	Have a slower rate of growth and there are signs of basic services being provided. Self-help schemes are often important, providing much needed improved housing, water and sewage disposal services. However, many people still work in the informal sector. An example is **Mumbai**, which has an annual population growth rate of 2.3%, 55% of the population living in slums, and 87% of houses having direct access to a piped water supply.	Have a more developed formal economy with large service-based industries. The majority live in legal, well-built homes and work in the formal economy. There are advanced transport, education and waste systems in place, and their governance is efficient. An example is **São Paulo**, which has an annual population growth rate of 1.9%, 30% of the population living in slums, and 94% of houses having direct access to a piped water supply.	Have stable and effective governance. They are often engaged in regeneration and urban sustainability projects. Many people work in highly paid professional service sector jobs, and hence their quality of life is high. An example is **London**, which has an annual population growth rate of 1%, and 100% of its houses having direct access to a piped water supply.

World cities

World cities also often have large populations but they are functionally very different from mega-cities. Whereas mega-cities have regional or national influence, the reach of world cities is global; they have become the command and control centres of the international economy. World cities dominate the global urban hierarchy and unlike mega-cities are mainly located in the developed world. The four top-ranking world cities are New York, London, Paris and Tokyo. Other 'rising stars' include Singapore, Hong Kong, Seoul and Mumbai, all in Asia.

Typically, world cities are leading business centres and the preferred headquarter locations of leading TNCs. They are also global service centres, specialising in advanced producer services such as finance, banking, accounting, management consultancy, law and advertising. In order to deliver these services to global markets, world cities are major telecoms, information and transport hubs. They are also magnets for highly educated, skilled workers, and home to world-class universities. Finally, world cities often have a political and cultural dimension: they house foreign embassies, consulates and international organisations, host international sporting events, and support a wide range of performing arts venues as well as museums, galleries and restaurants.

An alternative way to measure a World City is by its 'connectedness'. To be a world city, it is suggested that a city needs good transport networks to tie it into the world economy: a major international airport(s), and ideally its own docks. There should also be home-grown media and communications industries together with high-value jobs in international corporations, for example in financial services. Various indices, developed by international brand agencies such as Kearney, exist to identify such 'connectedness', such as the Global Power City Index and the Global Cities Index. The implication is that strong performance on some criteria leads to strong performance on the others: when a city becomes a global destination for, say, finance, it is more likely to become a cultural and political hub too. For example, Broadway in New York and Hollywood in Los Angeles are known throughout the world for their entertainment industries. In the developing world, there are equivalents in India – Bollywood in Mumbai, and Nollywood in Lagos, Nigeria – that are also becoming highly influential in the world of entertainment.

World cities are also important political centres and often host international conferences. Perhaps the largest is New York, which is the home of the United Nations. Some world cities are not that large in terms of population size, but still have a major influence on the world. One such example is Milan – the centre of the Italian and world fashion industry.

Urban forms

Over the last 100 years, sociologists, economists and geographers have developed a number of theories (or models) that seek to explain the form of urban areas (**urban form**). They have considered a range of physical and human factors that have shaped urban areas and have attempted to describe the spatial pattern of land use, economic inequality, social segregation and cultural diversity within them.

Knowledge check 25

Name some well-known, and some less well-known, mega-cities.

World city A city which has 'global' influence as a major centre for finance, trade, politics and culture. New York, London and Tokyo are pre-eminent.

Making connections

World cities are an integral part of several Global systems, for example economic, political and cultural.

Exam tip

Mega-cities and world cities are often confused. Make sure you know the difference between the terms.

Knowledge check 26

How can you measure a world city's brand?

Urban form The arrangement of land use in urban areas, its shape and spatial pattern.

Urban form in the developed world

In historical times, physical factors were the most important factors. Towns were built alongside rivers, on flat land, often around a river crossing (at a bridging point), or at the start of an estuary. As time progressed, economic factors involving land values became more important – the **PLVI** being crucial (Figure 4). The PLVI became the economic core of the urban area – the CBD – with other land uses (industry and residential) occurring as determined by the price of land as distance increases from the CBD (Bid rent theory – Figure 5). In recent decades, suburban office and shopping areas have resulted in secondary land value peaks within the bid rent landscape.

PLVI Peak land value intersection.

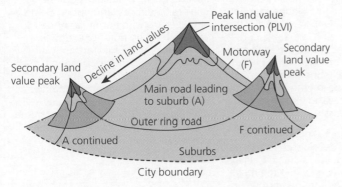

Figure 4 Land values for a typical city in a developed country

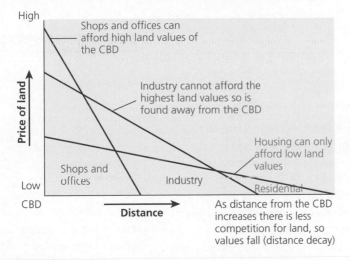

Figure 5 Bid rent theory

Based on these factors, four major models have been formulated, each seeking to explain where different types of people and businesses tend to exist within the urban setting. They are:

- The concentric zone model (Burgess)
- The sector model (Hoyt)
- The multiple nuclei model (Harris and Ullman)
- The urban realms model and 'edge cities'

The first three models were all based on Chicago, and were formulated in the early part of the twentieth century (Table 9). Other than Lake Michigan to the east, few physical features have inhibited Chicago's growth. Chicago includes a CBD, known as the Loop, because of the elevated railway lines around it. Residential suburbs go to the north, west, and south.

Table 9 Summary of three Chicago-based models of urban form

Concentric zone (Burgess)	Sector (Hoyt)	Multiple nuclei (Harris and Ullman)
A city grows outward, beginning with the CBD in the middle. Around this is the zone in transition where industry and poorer-quality housing can be found. Usually new immigrants to the city live in this zone in small quarters or rooming houses. The next ring is the zone of independent workers' homes. These are modest older houses occupied by the working class (or blue-collar workers). The next ring is the zone of better residences, where there are more spacious houses for middle-class families. Finally, the fifth zone is the commuter zone. People (often white-collar workers) who work in the centre, but choose to live in the suburbs.	A city develops in a series of sectors, not rings. Certain areas are more attractive for different activities. In the centre is the CBD. As the city grows, activities expand in a wedge, or sector, from the centre. Once a district with 'high-class' housing is established, the most expensive houses are built on the outer edge of that district further from the centre. Industrial and retailing activities develop in other sectors, as well as low-class and middle-class residential sectors. For Chicago, Hoyt argued that the best housing is developed north from the CBD along Lake Michigan, while industry was located along the major rail lines and roads to the northwest and southwest.	A city includes more than one centre (or nodes) that activities revolve around. Examples of these nodes include ports, a business centre, university, airport, and parks. Some activities go with particular nodes while others do not. For example, a university node may attract well-educated residents, bookshops, and publishers. Alternatively, the airport may attract hotels and warehouses. Likewise, incompatible land-use activities will not be clustered together. For example, industries will not be placed near high-class housing.

The urban realms model

This model is more recent, and better reflects the urban structure of large conurbations where a number of urban areas exist in close proximity. Examples of such areas include the Eastern Seaboard of the USA, London, and the urban area that includes Manchester and Liverpool in the northwest of England.

Such urban areas are linked together within a greater metropolitan region. In the early postwar period (the 1950s), rapid population diffusion to the outer suburbs created distant nuclei, but also reduced the volume and level of interaction between the central city and these emerging suburban cities. By the 1970s, outer cities were becoming increasingly independent of the original CBD to which these former suburbs had once been closely tied – i.e. they had their own 'realms'. For example, regional shopping centres (called malls in the USA) in the suburban zone were becoming the new CBDs of the outer nuclei. In the UK, these included out-of-town shopping areas on the edge of big towns and cities.

In 1991, the journalist Joel Garreau said that these outer cities at major suburban freeway (motorway) interchanges were the latest transformation of how people live and work – he called them 'edge cities'. The suburban cities (or suburban 'downtowns') at these interchanges have become home to office blocks and huge retail complexes (malls). Garreau states that we have moved our means of creating wealth, 'the essence of urbanism' – our jobs – to where most of us have lived and shopped for two generations (see also page 59).

Knowledge check 27

Choose one city you have studied and describe the social and economic inequality that exists there.

Exam tip

Consider how the growth and development of one named urban area you study or know reflects one or more of the urban models outlined in Table 9.

Urban form in the developing world

Most of the work that has been done on cities in the developing world has been based on South American cities. In 1980, Ford developed a model to describe the structure of cities in Latin America. This model states that Latin American cities are built up around a central core (or CBD). From that district comes a commercial spine that is surrounded by elite housing. These areas are then surrounded by three concentric zones of housing that decrease in quality as one moves away from the core (Figure 6).

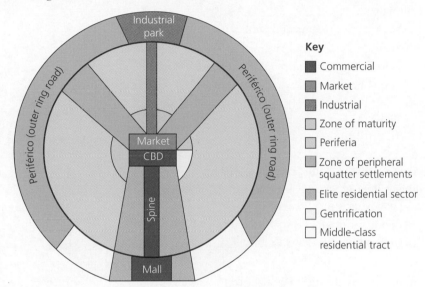

Figure 6 Ford's model of Latin American cities

Other features of the model include:
- a separate industrial sector/park develops on the edge of the city
- middle-class housing locates close to the elite housing sector and the **periférico**
- squatter settlements (**favelas**), fed by migrants, progressively develop in the **periferia**
- over time these favelas become more stable and obtain more services

Further revisions of the model have taken place:
- The central city should be divided into a CBD and a market area – this change shows that many cities now have offices, hotels and retail structures in their downtowns as well as their original CBDs.
- The spine and elite residential sector now has a mall or edge city to provide goods and services to those in the elite residential sector.
- Malls, edge cities and industrial parks are connected by the periférico, or ring highway, so that residents and workers can travel between them more easily.
- Some Latin American cities are also undergoing gentrification to protect historical landscapes – these areas are often located in the zone of maturity near the CBD and the elite sector.

Periférico An outer ring road around a Latin American city.

Periferia The outer parts of a Latin American city.

Favela A shanty town in Latin American cities.

Exam tip

Examine briefly how one named urban area in Latin America reflects Ford's model.

New urban landscapes

Town centre mixed developments

Most UK city centres have experienced economic decline in the last 30 years, caused by the development of out-of-town shopping retail parks and the decentralisation of many offices. High parking costs and congestion have assisted in this process. More recently, online shopping has added to the pressures that city centres face.

Consequently, many urban areas are encouraging the development of activities other than retailing. These include mixing retailing with:

- a range of leisure facilities – cinemas, theatres, restaurants and wine bars
- the provision of public gardens and squares
- the promotion of street entertainment, as in Covent Garden, London, and 'street art'
- new office, conference facilities and boutique hotels
- the construction of upmarket residential areas – often involving gentrification of old buildings such as old warehouses (see below)

Cultural and heritage quarters

These often form part of larger-scale redevelopment schemes, involving either:

- the presence of a distinctive cultural identity or product, e.g. a Chinatown, media industries, the arts, Victorian industrial goods, or
- a strong historical connection with an identity or product – such as a former dockland area (e.g. Bristol), or an old industrial product such as jewellery (e.g. Birmingham)

In some urban areas, planners have attempted to create such a quarter from scratch – building the buildings and infrastructure to merit the term of cultural quarter. The South Bank in London is one such example at a large scale; smaller towns such as Doncaster are also seeking to establish new cultural quarters.

Fortress developments

A recent trend in urban design is one of 'defensible space' that creates 'fortress developments' at a variety of scales:

- at a micro level these include 'anti-homeless' spikes or studs fixed into the ground outside private apartment blocks and hotels, and 'mosquito' alarms which emit a high-pitched sound to discourage loitering
- sloped bus shelter seats and park benches on which you cannot lie down or skateboard
- security systems such as high-perimeter fencing, barbed wire and CCTV cameras grafted on to public buildings, including schools and hospitals
- gated communities where access is achieved by key-pad

Gentrified areas

Gentrification involves the rehabilitation of old houses and streets on an individual basis, but is openly encouraged by groups such as estate agents, building societies and local authorities. The purchasing power of the new residents is greater, which leads to a rise in the general level of prosperity, and there is an increase in the number

Gentrification
A process of housing improvement by people rather than by organisations. It is associated with a change in neighbourhood composition in which low-income groups are displaced by more affluent people, usually in professional or managerial occupations.

of bars, restaurants and other higher-status services (Table 10). The very nature of the refurbishment that takes place in each house leads to the creation of local employment in areas such as design, building work, furnishings and decoration. There are several well-documented areas of gentrification in Britain's cities – Notting Hill, Islington and Camden Town in London, and Jericho in Oxford.

Table 10 Costs and benefits of gentrification

Costs	Benefits
People on low incomes cannot afford higher property prices or rents – this may impact on some ethnic groups more than others	Rise in general level of prosperity and increasing number and range of services and businesses
Higher car ownership may increase congestion	Increased local tax income for the local authority
Potential loss of business for traditional local low order shops	Physical environment of the area improved
Gentrifiers may be seen as a threat to the traditional community and friction may occur between 'newcomers' and original residents	Greater employment opportunities created in areas such as design, building and refurbishment

Edge cities

(Also see page 56.)

These have developed near to suburban freeway/motorway junctions and airports. For a place to be considered an edge city:

- the area must have substantial office space and retail space (such as a large shopping mall)
- the population of the area must increase every morning and decrease every afternoon (i.e. there are more jobs than homes)
- the place must be a 'single end destination' – with entertainment, shopping and recreation
- the area must have been previously undeveloped

The concept of the post-modern western city

The term 'post-modernism' refers to the changes that took place in western society and culture in the late twentieth century. It mainly concerned art and architecture, marking a move away from conformity and uniformity (modernism). Consequently, buildings in post-modern cities are often:

- more varied in design – curved, triangular, peaked, with greater amounts of exterior ornamentation
- the centrepieces of artistic and cultural quarters
- designed around the community that live and work there – giving more power and control to locals rather than being imposed by central government

Post-modern western cities are also stated to have greater ethnic diversities, with greater emphasis on cultural and artisanal industries and services.

Knowledge check 28

Give some examples of post-modern developments in cities.

Social and economic issues associated with urbanisation

Some of the major social and economic issues associated with urbanisation are those linked to **economic inequality**, **social segregation** and **cultural diversity**.

In general terms:

■ Economic inequality exists in terms of access to job opportunities, education, housing and basic services such as water and sanitation.

■ Knock-on impacts of these are poorer health, higher unemployment and a lack of social mobility for those at the poorest end of the spectrum – they get stuck in a cycle of poverty from which it is hard to escape.

■ Ethnic communities become isolated from wider society as they maintain their cultural diversity (origin, language, beliefs) and have limited interaction with others; or they could be perceived as outsiders by others in the society.

■ Consequently, both economic inequality and cultural diversity can lead to social segregation.

Social and economic segregation in the UK

Examples of social and economic segregation in the UK include:

■ when **gated communities** can be found adjacent to '**sink estates**', e.g. in the London Docklands

■ in rural areas, successful, prosperous commuter villages might be only a few miles from less attractive rural villages suffering population decline and service deprivation

■ expensive riverside property in London bought by wealthy European and Arab immigrants

■ wealthier white British people tend to live in leafy, suburban wards

■ lower-income ethnic groups are concentrated in areas with a large amount of social (council) or rented housing. For example, in Glasgow, asylum seekers and refugees are often placed in council-owned high-rise flats which would otherwise stand empty

There are other reasons why people from ethnic groups tend to cluster – Table 11.

Table 11 Factors behind ethnic clustering

Actions and attitudes of the ethnic group	Actions and attitudes of the rest of society
New immigrants tend to live close to already settled people from the same ethnic group, because they share a common language and experiences.	Estate agents or council housing officers may (consciously or unconsciously) help concentrate groups in particular areas.
Ethnically specific services – shops, places of worship, faith schools – encourage others to live nearby for convenience.	An existing population may leave an area if a new ethnic group begins to move in, making more housing available.
A view of 'safety in numbers' and stronger community ties if people live close together.	Prejudice (e.g. in jobs) prevents some ethnic groups gaining higher incomes to enable them to move away.

Economic inequality
The difference in levels of income and living standards in an area.

Social segregation
When different groups of people live apart from each other based on wealth, ethnicity, religion or age.

Cultural diversity
Variations in ethnicity or cultural values within a society.

Exam tip

It is important to recognise that all of these issues are interconnected – they don't exist in isolation.

Gated communities
Wealthy residential areas that are fenced off and have security gates and entry systems.

Sink estates Social housing estates that are the least desirable to live in. They tend to house the lowest-income and most in-need residents.

Social and economic segregation in the developing world

In the developing world:

- The consequences of inequality and cultural diversity are easily identifiable – the city centre core is the centre of wealth, and the periphery is the area of poverty.
- There are pockets of wealth dotted in other parts of a city, often 'gated' as in the developed world.
- It is estimated that nearly one billion people live in slums or favelas around the world, and this number is increasing – this will deepen the social, economic, political, and spatial inequality that is already prevalent.

The situation for people in slums, favelas, and shanty towns in the developing world is particularly harsh. As shown in the section on urban form, a core–periphery pattern exists within such cities and the various consequences of inequality are readily identifiable. For example, the favela Rocinha can be viewed as the periphery to core of Rio de Janeiro, Dharavi slum is the periphery to central Mumbai, and the Kibera slum is the periphery to central Nairobi.

Two important features of inequality within urban areas in developing world countries are worth emphasising:

- The spatial embedding of a growing urban underclass in periphery-based slums, favelas, and shanty towns.
- The perception of individuals and communities located within periphery spaces by those in the urban core as a component of their own identity. This can be illustrated by the use of the expression 'Favelado' to disparagingly describe those who live in the favelas of Brazil's major cities.

Favelas as places

The urban space of the slum, favela, or shanty town has been thought of by some as an outsider space: a space of lawlessness, and a space of the undesirable poor. These spaces not only point to large-scale systemic problems (such as economic failure or overpopulation), they also shed light on the way state and local authorities create and exacerbate the conditions of separation that impact the lives of those who inhabit these areas. To these agencies, slums can represent a kind of built environment that incorporates long-standing prejudices and historical inequities.

An illustration of this can be seen in the city of Rio de Janeiro, Brazil, which has long struggled with its own version of the core–periphery relationship as the city tries to manage life in and around its estimated 600 favelas. As a means of combating the favelisation of the city, there are plans to construct walls around some of the city's larger favelas (e.g. Rocinha) as a means of halting the growth of these improvised housing settlements. Among the various forms of social and spatial control developed to combat favelisation is the euphemistically named 'eco-barrier' which is intended to protect the natural beauty of rolling green hills that surround the city from the unauthorised development of ever-growing favelas. A secondary effect of this barrier is to isolate and conceal the manifest signs of economic inequality that are associated with Rio's growing favela communities. This type of response by local government may reflect a social system that favours an exclusive, elite vision of urban growth in developing world cities.

Exam tip

Consider some of these factors in a named area you know where there is ethnic segregation.

Making connections

This section clearly links to insider and outsider perspectives within Changing places.

Strategies to manage these issues

Various governments (national and local), NGOs and individuals (e.g. the Gates Foundation) seek to manage these issues. In general, this involves:

- managing socioeconomic issues, e.g. improved provision of schools, enforcing a living wage, giving access to affordable housing, greater provision of public transport
- dealing with social variations, e.g. increasing the availability of clinics; health education programmes involving access to sports and leisure facilities
- reducing segregation by legislation on anti-racism and employment rights; combating discrimination and prejudice; encouraging greater political involvement of different cultural groups
- dealing with issues of cultural diversity, e.g. providing language lessons or bilingual literature; hospitals catering for specific illnesses; schools altering their curricula and holiday patterns to cater for different ethnic groups

Finally, you should examine all of these issues and the strategies to manage them in the context of the two case studies of contrasting urban areas you have to undertake. Make sure you look at patterns of economic and social wellbeing, and the impact of environmental conditions for these two places. Also ensure that you examine strategies that deal with both economic inequality and cultural diversity.

Urban climate

The impact of urban forms and processes on local climate and weather

Urban climates provide a local-scale example of the impact of human activity on the atmosphere. Cities create their own climate and weather. Some geographers refer to this as the 'climatic dome' within which the weather is different from the surrounding rural areas in terms of temperature, relative humidity, precipitation, visibility and wind speed. For a large city, this dome may extend upwards for up to 250–300 m and its influence may well extend tens of kilometres downwind.

Within the urban dome, two levels can be recognised. Below roof level there is an urban canopy with processes acting in the space between buildings. Above this, there is the urban boundary layer, whose characteristics are governed by the nature of the urban surface. The dome extends downwind as a plume into the surrounding rural areas and this phenomenon can occur at height, but with its effects being absent at ground level.

Urban temperatures: the urban heat island effect (UHIE)

The **urban heat island** effect (UHIE) (Figure 7) is the product of a variety of factors. These include:

- anthropogenic heat sources that include warmth given off by people, machines, heating systems, air conditioning systems, industrial processes and cars
- multiple reflections of **insolation** from tall buildings, especially those with high levels of glass
- urban surfaces (concrete, bricks, tarmac) tend to have a lower **albedo** which enables them to absorb more of the incoming solar radiation; the higher heat capacity of urban surfaces allows them not only to absorb the heat but also to store it

Urban climate A set of climatic conditions that prevail in a large metropolitan area, which differ from the climate of its rural surroundings.

Making connections

There are some connections to the Water cycle here, specifically the factors affecting the hydrological cycle in urban areas.

Urban heat island The zone around and above an urban area which has higher temperatures than the surrounding rural areas.

Insolation Incoming solar radiation.

Albedo The amount of solar radiation that is reflected by the Earth's surface and the atmosphere.

- this heat is then released slowly when the air cools at night
- efficient drainage of the urban surface removes surface water quickly – there is less capacity for evaporation to take place, with its associated cooling effect
- there is less vegetation, which would cool the air by transpiration
- above many cities there is a dome of particulate and NO_2 pollution – this allows the short-wave radiation from the sun into the atmosphere but then absorbs and reflects the outgoing longer-wave radiation, preventing its escape
- often increased cloud amount over the urban area also reflects outgoing radiation back to the surface

Knowledge check 29

The UHIE often occurs when there is a temperature inversion. What is a temperature inversion?

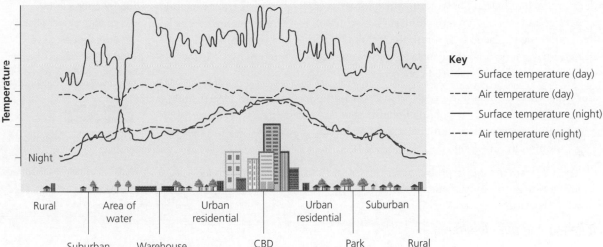

Key
—— Surface temperature (day)
- - - Air temperature (day)
—— Surface temperature (night)
- - - Air temperature (night)

Figure 7 The urban heat island effect

The UHIE develops best under certain meteorological conditions. The contrast between urban and rural areas is greatest under calm, high pressure (anticyclonic) conditions, particularly with a **temperature inversion**. Heat islands are often well developed in winter when there is a bigger impact from city heating systems and they are much more distinct at night when the impact of insolation is absent and surfaces, having absorbed heat by day, can slowly release it back into the atmosphere after the sun has gone down.

The UHIE is also spatially variable (Figure 7). The edge of the island is usually well defined as temperatures change abruptly from urban to rural. Some climatologists have even likened the effect to a 'cliff' in temperatures. From here, the temperature rises steadily to a 'peak' in the city centre where building densities are highest. Micro-hotspots may also appear associated with large car parks, shopping centres and industrial premises. Equally there may be micro-cold spots linked to features such as parks, fields and water bodies.

Increasingly, scientists are studying the UHIE in summers when potentially lethal heatwaves can occur. The UHIE is exacerbated during hot spells, as it reinforces air temperature increases particularly in poorly ventilated outdoor spaces or inner spaces in residential and commercial buildings with poor thermal insulation. This in turn increases energy consumption for cooling machinery – air conditioning and

Temperature inversion
An atmospheric condition in which temperature increases with height rather than the more usual decrease.

refrigeration units – which creates more demand for power from power plants. All of these add to costs of energy production as well more greenhouse gases being emitted at power plants. Furthermore, a direct relationship between UHIE intensity peaks and heat-related illnesses and even fatalities has been found due to the incidence of thermal discomfort on the human cardiovascular and respiratory systems. Heatstroke, heat exhaustion and heat cramps have increased in incidence, especially in the elderly and children.

Dealing with the UHIE is something which urban planners are just beginning to address. One suggestion is to increase surface reflectivity (i.e. to increase the albedo) in order to reduce radiation absorption by urban surfaces. Reflective surfaces include having light colours, or even white paint, on construction material. In some cities a white membrane is placed over a building's roof – called a 'cool roof'. Similarly, whitening of pavements and road surfaces has been used – 'whitetopping' – using white rock chips or lighter-coloured concrete.

Another suggestion is to increase vegetation cover in the form of urban forests and parks in order to maximise the cooling benefits that vegetation has. At a smaller scale, 'green roofs' and 'green walls' can increase cooling by transpiration – for example, these are common in Singapore.

Finally, the UHIE could cause problems in the future. With little access to air conditioning, refrigeration, or medical care, the world's urban poor will be particularly vulnerable to heatwave-related health hazards.

Precipitation and thunderstorms

Urban areas have up to 10% more rainfall than surrounding areas. Rain lasts for longer periods and intense storms with thunder are more common.

The increase in rainfall is because:

- the UHIE causes relatively low atmospheric pressure and convectional uplift and consequent convectional rainfall, which also increases the number of thunderstorms and the intensity of the rainfall
- the presence of high-rise buildings can cause turbulence which can lead to uplift of air
- particulate pollution means that there is an increased number of **hygroscopic** nuclei present in urban air

There are also reduced amounts and incidents of snowfall in urban areas due to the warmer temperatures, though when they do occur they tend to be highly disruptive of urban transport systems.

Fogs

Relative humidity is lower in cities than the surrounding rural areas (due to fewer water bodies, lower rate of evapotranspiration and more rapid runoff of water). At night though, an urban area maintains its humidity, whereas in rural areas the air cools more rapidly and moisture is lost to dewfall. There is also a greater concentration of airborne particulates in urban areas (see air quality below) that act as condensation nuclei. These factors, together with the higher levels of pollution in cities, mean that there is a much higher incidence of fog in urban areas, particularly under anticyclonic conditions.

Making connections

There are some connections to the Carbon cycle (emission of greenhouse gases) and the impacts of future climate change here.

Exam tip

The UHIE is a significant concept – it might help to learn it in terms of causes, consequences and solutions.

Hygroscopic Water attracting.

Fogs are much more frequent in winter and 30% more frequent in summer than in surrounding rural areas. Urban fogs are most common near rivers. These are often artificially warmed by effluent from industry and sewage works, evidenced by rising steam, resulting in increased levels of water vapour in the atmosphere.

Urban winds

The surface area of cities is uneven due to the varying height of the buildings. Buildings tend to exert a powerful frictional drag on air moving over and around them. This creates turbulence – rapid and abrupt changes in both wind direction and speed. Average wind speeds are lower in cities than the surrounding areas and they are also lower in city centres than in suburbs. In urban areas:

- Annual mean wind speeds are 20–30% lower.
- The frequency of calms can be 5–20% higher.
- High-rise buildings may slow down air movement but they also channel air into the 'canyons' between them (the Venturi effect). Winds in the canyons between buildings can be so powerful that they make buildings sway, blow over vehicles and knock pedestrians off their feet.
- Hence, gusts of winds can often be much stronger in central urban areas.

A single building can modify an airflow passing over it:

- Air is displaced upwards and around the sides of the building and is also pushed downwards in the lee of the structure.
- On the windward side, air pushes against the building creating relatively high pressure – this increases with height, causes a descending flow which forms a vortex when it reaches the ground and sweeps around the windward corners.
- Air flowing around the sides of the building can become separated from the walls and roof and create suction in these areas.

A group of buildings:

- creates a disturbance to the airflow which depends upon the height of the buildings and the spacing between them
- if they are widely spaced, each building will act as an isolated block (as above)
- if they are closer, the wake of each building interferes with the airflow around the next structure and produces a complex pattern of airflow that is difficult to predict

Air quality

Air quality is a direct reflection of the extent to which there is atmospheric pollution in urban areas. The amount of air pollution in the atmosphere depends on the rate at which pollutants are produced and the rate at which they are dispersed (diluted) as they move away from their source. The key atmospheric pollutants that are likely to have an impact on health are: ozone (O_3), nitrogen dioxide (NO_2), sulphur dioxide (SO_2) and particulate matter from vehicle exhausts, cement dust, tobacco smoke and ash (PM_{10}) and fine particulate matter ($PM_{2.5}$).

Concentrations vary in time and space, because of variability in the sizes and locations of polluting sources. For example, in most cities a significant local source of air pollution is traffic emissions. These vary significantly depending on factors including:

- the volume and speed of traffic
- the types and proportions of vehicles (buses, lorries, cars, motorbikes etc.)

> **Exam tip**
>
> Be aware of some locations within named urban areas where variable wind speeds have caused issues – for example Bridgewater Place, Leeds.

- the age and level of maintenance of vehicles
- the temperature of the engine and fuel type used

The geography of the urban built environment (such as the height of buildings and how close they are to the road, the width of the road and shape of the road network, local topography and presence of vegetation) can also affect air pollution concentrations. This means that the highest concentrations are not only found where the emissions are highest, but also in areas where pollutants get trapped and do not disperse effectively. Such pollutant hotspots move around over time – a hotspot observed during morning rush hour may not be found at the same place during the evening rush hour.

Photochemical **smog** is another source of poor air quality. This is the result of a chemical reaction between sunlight and nitrogen oxides (NO_x) and leads to high levels of ozone in the lower atmosphere. This can be the cause of health problems (headaches, eye irritation, coughs and chest pains) as well as being damaging to vegetation. Los Angeles has had a serious problem with photochemical smog because of its high density of vehicles, frequent sunshine and the favourable topography that traps the high concentration of photo-oxidant gases at low levels.

Smog A mixture of smoke and fog exacerbated by high levels of industrial and vehicular pollution.

Pollution reduction policies

There are a number of ways in which governments (local and national) have tried to reduce atmospheric pollution in cities.

- **Atmosphere measures:** Clean Air Acts, smoke-free zones and regulations on levels of airborne pollution, particularly on the level of PM_{10} particulates in the atmosphere.
- **Vehicle control:** traffic-free zones, pedestrianised areas, congestion charges and restrictions by registration plate numbers – odd numbers one day, even numbers the next.
- **Public transport:** persuading more people to use public transport rather than bring their cars into the city. Such schemes have included the development of tram systems, bus-only lanes and park-and-ride schemes. Several cities have bike loaning schemes.
- **Zoning of industry:** placing industry downwind in a city; planning legislation has forced companies to build higher factory chimneys that emit pollutants above the dome layer.
- **Vehicle emissions legislation:** encouraging manufacturers to develop better fuel-burning engines and introduce catalytic converters to remove most of the particulates from exhaust fumes. Hybrid and electric cars should also have an impact in the future.

Knowledge check 30

Outline the vehicle restrictions that London has put in place to address air pollution.

Urban drainage

Urban catchment issues

Urban areas have a modified hydrological system, or cycle (Figure 8). In general, towns and cities comprise larger proportions of impervious surfaces (roofing materials, concrete, tarmac, paved driveways) and are generally constructed on flat or gently sloping areas with limited natural drainage opportunities. Hence infiltration is much reduced. Impermeable roofs and roads are shaped to get rid of water quickly and combined with a dense network of drains and sewers, this means that water gets to an urban river very quickly, reducing lag time and increasing discharge. Hydrographs in urban areas therefore are more flashy (steeper rising and recessional limbs) and have higher peak discharges. However, it must be pointed out that urban areas simply exacerbate prevailing weather conditions over an area. Urban areas often flood because they try to control the water passing through them that has fallen onto or arisen from upland areas upstream.

Making connections

There are obvious connections to the Water cycle here, specifically the factors affecting the hydrological cycle within urban areas.

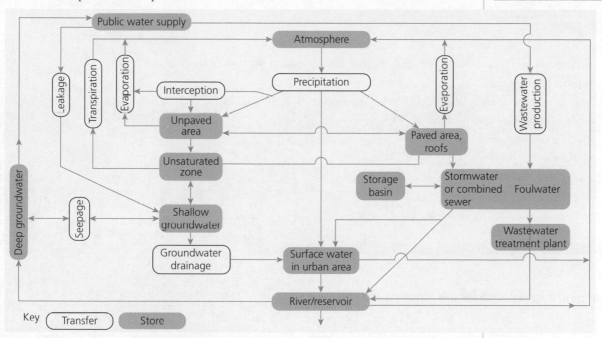

Figure 8 The urban hydrological cycle

Surface water flooding in urban areas is becoming an increasingly common phenomenon, and it is thought that it will become more likely with climate change. Intense rainfall on such surfaces readily ponds up before it can escape via man-made drainage systems. These are sometimes inadequate, or become blocked by vegetation and/or litter (wet wipes are notorious), and hence flooding is quickly generated. For example, during the 2008 Morpeth floods, huge amounts of debris blocked culverts upstream of the town, which had been constructed to allow water to drain quickly through the town. In extreme cases, this can lead to some 'throttling' of flow in surface water drains and sewers and actually slow down movement to the river thereby reducing the potential peak, to the cost of the areas already inundated. The inability to cater for such increased amounts of surface water efficiently means that in periods of intense rainfall, water will flow either across the surface in large volumes or through subsurface sewerage systems, and in some cases may even cause the latter to burst back out on to the surface through 'popped' drain covers.

Where man-made drainage systems are effective, surface water is channelled directly into drains and sewers in an urban area, so precipitation reaches the river quickly. However, this effect is likely to be more damaging to areas further downstream. River channels may also become constricted by bridges, which can slow down discharge and reduce the carrying capacity of the river. In times of spate, debris can be deposited directly behind bridge supports and exaggerate the effects of a flood.

Sustainable urban drainage systems (SuDS)

SuDS seek to increase rates of infiltration, thereby reducing surface water flooding and simultaneously improving the quality of water that eventually emerges into rivers.

SuDS include:

1 Constructed facilities and materials to store and/or drain water naturally:
   - permeable surfaces such as porous linings around trees and grassed areas
   - infiltration trenches
   - rainwater harvesting – ponds and swales (artificially created hollows)
   - underground storage
   - green roofs and walls
   - wetlands
2 Practices that involve management of water quality
   - mitigation of pollution accidents
   - reduction of polluting activities, e.g. bunding of oil tanks

River restoration/conservation example: the Olympic Park, Stratford, London

Background and reasons for the project

The Olympic Park development in Stratford, east London, covered an area of approximately 250 ha and housed the Olympic (now London) Stadium, Aquatics Centre, Velodrome, Copper Box, BMX Track, as well as extensive public access areas (landscaped and paved), transport centres and other operational facilities.

The original site was formerly industrial/commercial development together with Lea Valley Park and was a known depository for building rubble from properties demolished during the Second World War. The site was known to be contaminated throughout and overlays deposits of river terrace gravel and chalk aquifers. Even after completion of the various works and remediation, the site still contains significant areas of residual contamination. This limited the opportunity for employing infiltration drainage systems across the site. A number of strategic watercourses traverse the Park including the River Lea, River Lee Navigation, Waterworks River, City Mill River and Old River Lea.

The Park is designed to be protected against river flooding within a 100-year return period rainfall event. The site topography was dramatically changed during the course of the development, with some areas of the Park being raised by 9 m. In addition, the creation of the wetland bowl within the River Lea required significant widening of the river channel, reducing existing land levels around it. Plateaus have been formed for the sporting venues and associated facilities above the river flood level with access routes to the level of the towpaths adjacent to the watercourses.

Knowledge check 31

Identify three differences between an urban storm hydrograph and a rural one.

Sustainable urban drainage systems (SuDS) An approach to managing rainfall in urban areas that seeks to replicate natural drainage systems, managing it close to where it falls.

SuDS used

A series of SuDS were used in the project:

- Porous asphalt strips are extensively employed throughout the pedestrian concourses of the Park. These act as collection points for any runoff generated and convey water into trenches below the strips which contain perforated pipes. The perforated pipes (made from recycled thermoplastic material) then drain to catch-pits and then into the rivers.
- Within the wetlands located adjacent to the Basketball Arena/Velodrome swales, filter strips/drains and small balancing ponds have been used.
- Swales are included within the wetlands area as slowing-the-flow conveyance devices. They incorporate check dams or weirs at intervals to reduce velocities and provide open water features (attenuation ponds) alongside wetland pathways.
- Rainwater harvesting was installed at two permanent venues – the Velodrome and Copper Box.
- Within the roadways of the Park, traditional road gullies and kerb drainage collection systems were used.

Evaluation

The SuDS designed and installed within the Park take account of the requirements of the UK Environment Agency:

- They take into account the potential impacts of climate change.
- They have reduced sources of surface water contamination and increased the use of appropriate methods of interception, infiltration and treatment.
- The opportunity to enhance the biodiversity and amenity of the public space – the North Park wetland feature has become a wildlife haven for plants and animals with habitats created for otters, kingfishers, grey herons and water voles.
- The use of appropriate materials in the construction and maintenance of the drainage infrastructure, much of which is recycled and carbon-neutral.
- The safe operation and maintenance of the drainage system through its operational life including the efficient transfer from Olympic Games' activities to Legacy uses.

Exam tip

You could consider a SuDS scheme closer to where you live – follow the same headings as this case study.

Urban waste and other environmental issues

Sources of waste

Main points:

- The average person in the UK produces over 500 kg of household waste per year.
- Waste is also generated in huge amounts by industrial and commercial activity.
- Medical institutions create a specific type of waste, which needs careful processing.
- A range of industries exist to manage such waste, some of which are international.
- For example, several plants exist in the UK that recycle parts from cars and motor vehicles – they are much more sophisticated than 'scrapyards'.
- Some trade in industrial waste is difficult to follow – an example is the shipment of electronic waste around the world.

■ Many people feel that the amount of waste generated is unsustainable and that we need to change our attitude to it – to see it as a resource to be managed rather than as a nuisance (reflected in the growing interest in the **circular economy**).

The relationship between waste and economic characteristics and lifestyles

Globally, waste increases by about 7% per year. Population growth accounts for much of this increase but economic development also plays a role, since greater personal wealth increases consumption of goods and services and this leads to more waste. In urban areas, where there are large concentrations of people, the amount of municipal solid waste (MSW) is particularly high and is set to increase significantly as a result of urbanisation and rising living standards.

Solid waste is very much an 'urban issue' because urban residents produce about twice as much waste as their rural counterparts. Globally, rural dwellers tend to be poorer, purchase fewer store-bought items (which results in less packaging), and have higher levels of reuse and recycling. Waste generation varies significantly between cities. Rates of waste production are much higher in cities in high-income countries as disposable incomes and living standards are higher. However, it is the cities in low and middle-income countries which are set to see the biggest increase in waste generation as a result of rapid urbanisation and continued industrialisation. The costs of collecting and treating waste are high. In lower-income countries, solid waste management is usually a city's single largest budgetary item and it is common for urban authorities to spend up to 50% of their budget on solid waste management. Much is placed in landfill sites or simply dumped in empty spaces or nearby countryside areas.

Finally, waste is a large source of methane, a powerful greenhouse gas (GHG). Waste also contributes to water, ground and air pollution. Untreated or uncollected waste can lead to health problems such as respiratory ailments, diarrhoea, cholera and dengue fever.

Waste disposal

See Table 12.

Table 12 Methods of waste disposal

Method	Commentary
Unregulated	Waste that is placed in illegal locations – any disposal site – which could include hot/dirty water into rivers; landfill is sometimes said to be unregulated.
Recycling	The reprocessing of waste – metals, plastics, paper – into new products; is now a major industrial activity in many countries both formally and informally; targets – by the EU, the UK and local authorities – for recycling are now common practice. Reducing the amount of packaging and plastic bags is a form of recycling.
Recovery	The reuse of organic materials, using digestive energy-producing plants or composting of organic waste; needs careful management as the by-products can be dangerous.

Circular economy
An economic system aimed at eliminating waste and supporting the continual use of resources.

Making connections

There are some connections to the Carbon cycle (emission of greenhouse gases) here.

Knowledge check 32

Describe the main features of the movement of electronic waste around the world.

Method	Commentary
Incineration	A form of energy recovery by burning waste; most sites in the UK use the heat created to warm buildings or generate electricity; there are some serious concerns regarding atmospheric pollution, denied by companies.
Burial (landfill)	The waste is buried – dumped in old quarries or hollows where it is unsightly and a threat to groundwater supplies and river quality as toxic chemicals are leached out; decomposing waste also emits methane, the most toxic of the GHGs and potentially explosive; to discourage use of landfill sites, landfill taxes are often imposed, and the activity is now closely regulated in the UK.
Submergence	The 'burial' of waste at sea is banned by international convention; illegal dumping of ship oil still takes place.
Trade	Electronic waste is shipped around the world, from the developed to the developing world; this trade in industrial waste is difficult to follow.

Making connections

There are some connections to world trade patterns (Global systems), and the global commons (Global governance) here.

An example of waste disposal: Sheffield, UK

The city of Sheffield is under ever-increasing pressure to manage and dispose of its waste in a more environmentally sustainable and cost-effective manner. The tightening of environmental legislation, including the EU Landfill and the EU Waste Incineration Directives, now means that the city can no longer continue to send a large proportion of its household waste to landfill sites, and this is now less than 15%.

Sheffield has an approach integrating a waste incinerator (called an Energy Recovery Facility, or ERF), located centrally, with a network of pressurised hot water pipelines under parts of the city centre (called the District Energy Network) to recover heat from household waste. The incinerator is run by Veolia Environmental Services. The majority of waste collected in the city is taken to the ERF where it is burnt at temperatures of over 850°C in a specially controlled environment. The heat created from the process is converted to steam and used to generate heat and electricity. The facility is designed to handle 225,000 tonnes of municipal solid waste a year. Up to 60 MW of heat is supplied to over 140 buildings connected to the District Energy Network including offices, leisure facilities (the Ponds Forge swimming pool), hotels, and new apartments. The number of buildings wanting to connect is continually growing.

The plant also generates up to 19 MW of electricity for the National Grid; enough to power 19,000 homes. Heat provided by the District Energy Network in Sheffield saves 12,000 tonnes of carbon emissions from being released into the atmosphere every year.

Exam tip

You could consider an urban waste disposal scheme closer to where you live.

Other environmental issues

Atmospheric pollution in urban areas has been addressed earlier (see page 65).

Water pollution

Sources

■ Runoff from streets carries oil, rubber, heavy metals, and other contaminants from cars and other vehicles.

■ Untreated/poorly treated sewage water can be low in dissolved oxygen and high in pollutants such as faecal coliform bacteria, nitrates and phosphorus.

- Treated sewage can still be high in nitrates.
- Warm water from power stations – thermal pollution.
- Groundwater and surface water can be contaminated from waste dumps, toxic waste and chemical storage areas, leaking fuel storage tanks, and intentional dumping of hazardous substances.
- Domestic water drains can carry cooking oil, paints, detergents and litter (e.g. wet wipes).
- Misconnections – misconnected properties let **greywater** enter surface water drains.
- One-off incidents such as a road tanker accident.

Consequences

- Oils, cooking oils and fats spread out across the top of a watercourse and cause a rainbow effect called iridescence.
- Larger amounts can create a matt effect on a water surface and pools of oil may weather and solidify.
- Greywater can cause algae and fungus to grow at outfalls and in watercourses.
- Polluting substances deplete oxygen from water, causing living things to 'suffocate' – they can also be directly toxic to animals, fish and plants.

Strategies

Solutions involve sustainable ways for an urban area to:

- reduce its dependence on pollutants and the amount of pollutants it produces
- recycle or dispose of pollutants before they contaminate water
- educate industries and individuals in ways to reduce polluting activities

Specific strategies include:

- treatment plants
- legislation and enforcement
- combining with SuDS to improve water quality naturally

Dereliction

Problems

- A negative perspective is created of an area – vandalism increases; house prices nearby fall.
- Land is often contaminated, and dangerous to trespassers.
- Areas are colonised by plants that can be challenging to remove, e.g. Japanese knotweed spreads easily via rhizomes and cut stems or crowns; it is now listed under the Wildlife and Countryside Act 1981 as a plant that is not to be planted or otherwise introduced into the wild.

Strategies

- Total clearance as part of urban regeneration schemes.
- New building developments on **brownfield sites**.
- Remediation – removal of pollutants and toxins.
- Community action, e.g. small-scale urban farms being established.

Greywater Polluted water originating from bathroom sinks, showers, tubs and washing machines.

Exam tip

Be aware of a water pollution issue and possible solutions to it where you live.

Dereliction Land that has been abandoned, such as after the clearance of former industrial sites, and then becomes dilapidated.

Brownfield site Previously developed land, now standing idle.

Knowledge check 33

Outline the remediation of one former derelict area you have studied.

Sustainable urban growth

Cities pose a threat to both the local and global environment:

- by creating pollution (air, water and dereliction) they threaten **sustainability**
- they can have a very high **ecological footprint**

Consequently, there have been pressures to enshrine sustainability in cities (Table 13).

Table 13 Dimensions of sustainability

Economic	Social	Environmental	Governance
Individuals and communities should have access to a reliable income over time. Employment opportunities for all.	All individuals should enjoy a reasonable quality of life, with access to education, food, health services, clean water and sanitation.	No lasting damage should be done to the environment. Renewable resources must be managed in ways that guarantee continued use.	Commitment to sustainable policies by decision-makers, such as 'green' planning policies. Strategies to reduce inequalities.

Sustainability Meeting the needs of today without compromising the ability of future generations to meet their own needs.

Ecological footprint A measurement of the area of land or water required to provide a person (or society) with the energy, food and other resources they consume and render the waste they produce harmless.

The thrust of 'sustainability' is to improve the 'needs of today' in a way that does not have to be paid for tomorrow. In applying this principle to urban areas, three questions can be asked:

1 What are the 'needs of today'?

They can be summarised as:

- minimising the ecological footprint of society
- improving the quality of the living environment
- waging war on deprivation and discrimination
- raising public participation in government and decision-making
- ensuring a sound economic base

2 Is it possible to meet these needs today?

The realistic answer must be 'no'. Urban areas throughout the world are, in varying proportions, failing on all or most of the five counts above.

3 Will it still be possible to meet these needs tomorrow?

Again, the most likely answer is 'no'. Urban areas will always, despite our best efforts, consume non-renewable resources, pollute the environment and embody risk.

Once there is a recognition that urban areas cannot, and will never, become wholly sustainable, attention can focus instead on ways to make our urban areas *more* sustainable than they are at present. A move in this direction depends on taking courses of action relating to the needs set out above.

Knowledge check 34

When and where did the concept of sustainability first appear?

Link with systems

It may be useful to set these possible actions in a context that sees the urban area as an open system, with inputs, outputs and internal actions. The inputs of an urban area are mainly made up of those resources that are brought in to support urban life. They range from food and water to building and industrial raw materials. The problem is that some of those resources are non-renewable and therefore exhaustible. The outputs of urban areas include waste, pollution and the spread of the built-up area into the countryside. The ecological footprint of an urban area is the sum total of its inputs and outputs, particularly the former. A reduction in this footprint is a big step in the direction of sustainability, though it needs all stakeholders (e.g. governments, TNCs and individuals) to agree and act.

In summary: to achieve sustainability, cities should:

- have less reliance on fossil fuels
- increase the use of technology to offset the use of fossil fuels
- exploit renewable energy
- have more sustainable water usage
- develop more sustainable transport systems
- encourage circular economies

Liveability

Linked to sustainability is the concept of **liveability**:

- A combination of economic, aesthetic and environmental factors provides a much more balanced perspective on a city's 'value' to its citizens.
- An index of liveability can be created using measures of economic vibrancy and competitiveness, domestic security and stability, socio-cultural conditions, public governance, environmental friendliness and sustainability.
- In 2019, the top five best cities to 'live in' were judged to be Vienna, Melbourne, Sydney, Osaka and Calgary.

Sustainable cities

Features

See Figure 9.

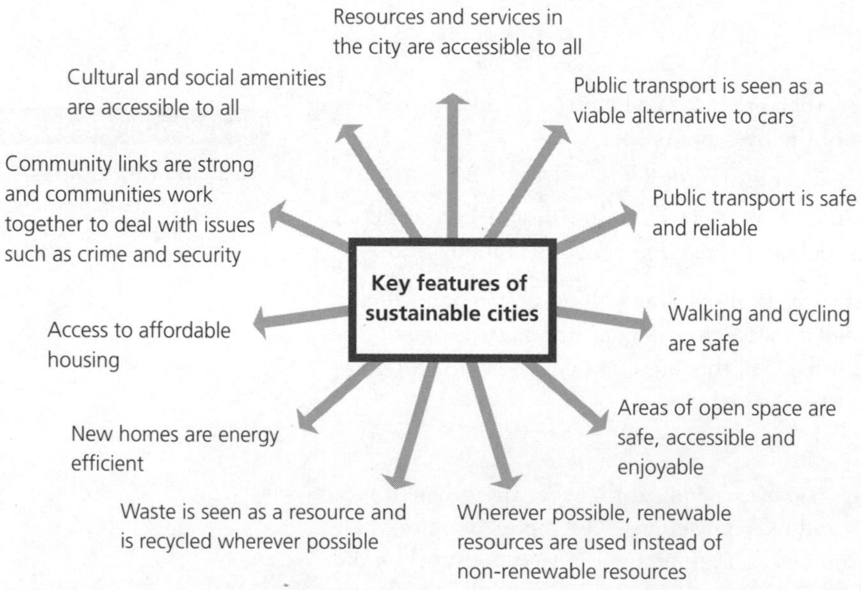

Figure 9 Features of sustainable cities

Making connections

Systems theory is a key concept that applies to both physical and human geography.

Liveability A measure of what it is like to live in a city and how urban life there compares with other cities in the world. It assesses 'work–life' balance, environmental awareness and a sense of localism versus globalism.

Exam tip

An annual survey of liveability is published every year – keep up to date at:
http://www.eiu.com/topic/liveability

Strategies for developing more sustainable cities

- Increasing the use of renewable sources of electricity.
- Appropriate systems of waste management, including dealing with food waste.
- Reducing fuel emissions within urban areas – buses, lorries, railways, airports.
- Investment in affordable and low-carbon housing.
- More mixed-use developments where people can live, work and relax in the same area.
- Establishing household goods recycling – investment in reuse schemes and provision of space for reuse activities.
- Reduced consumption of water – wastage through pipe leaks can be a huge problem.
- Greater use of public transport, e.g. the rollout of technologies such as hydrogen fuel cell buses and hybrid electric buses, electrification of the railway network and biodiesel for trains.
- Local sourcing – choosing the most local supplier for standard products.
- 'Green governance' – ensuring all decision-making is aimed at sustainable outcomes.

Knowledge check 35

Name some examples of cities that have been declared to be sustainable. In what ways are they more sustainable than other cities?

Case studies

You are required to study case studies of **two contrasting** urban areas to illustrate and analyse all of the key themes set out above, to include:
- patterns of economic and social wellbeing
- the nature and impact of physical environmental conditions

This must be done with particular reference to the implications for sustainability, the character of the study areas and the experience and attitudes of their populations.

Possible examples include: London, Liverpool, Johannesburg, Mumbai and Barcelona.

Summary

After studying this topic, you should be able to:
- understand the importance of urbanisation in human affairs, and be able to describe the global pattern of urbanisation, including the recent growth and development of mega-cities and world cities
- know and understand the various processes associated with urbanisation (including suburbanisation, counter-urbanisation, urban resurgence, deindustrialisation, decentralisation and regeneration)
- know and understand the variety of urban characteristics and forms that reflect spatial patterns of land use, economic inequality, social segregation and cultural diversity, in both the developed and developing world
- recognise and describe the variety of new urban landscapes that exist today, including heritage quarters, fortress developments, gentrified areas, edge cities and the 'post-modern' western city

- evaluate the issues associated with economic inequality, social segregation, and cultural diversity, including the strategies to manage them
- know and understand how the climate of urban areas varies from that of the regions immediately around them, including differences in temperature, rainfall, winds and air quality
- know and understand how urban drainage systems operate, and in particular the strategies associated with sustainable urban drainage systems, and river restoration and conservation schemes
- evaluate the issues associated with waste in urban areas and the means by which it can be disposed of, with particular reference to incineration and landfill
- evaluate other contemporary urban environmental issues including water pollution and dereliction, and the strategies to manage them
- know and understand the principles of sustainable urban development (including the concepts of ecological footprint and liveability), and be able to evaluate strategies for developing more sustainable cities

Population and the environment

Introduction

The environmental context for human population characteristics and change

Table 14 Environmental elements affecting sparsely populated areas

Environmental element	Sparsely populated areas
Relief	Rugged mountains where temperatures are low (the Andes, the Himalayas)
	Active volcanic areas (Iceland)
Climate	Areas of very low annual rainfall (Sahara desert)
	Areas of long seasonal drought (the Sahel)
	Areas with high humidity (the Amazon)
	Very cold areas (northern Canada, Siberia)
Vegetation	The coniferous forests of northern Eurasia and northern Canada
	The rainforests of the Amazon and the Congo basins
Soils	The frozen soils (permafrost) of the Arctic and Siberia
	The thin soils of mountainous areas
	The leached soils of the Amazon rainforest where forest clearance has occurred
	The overgrazed areas of the Sahel
Resources	Areas lacking in fuel resources and valuable mineral resources
	The areas where extensive farming takes place
Water supplies	Areas lacking a permanent supply of fresh clean water due to irregular rainfall, or few wells and reservoirs (Sudan and Ethiopia)

Table 15 Environmental elements affecting densely populated areas

Environmental element	Densely populated areas
Relief	Flat lowland areas (The Netherlands and Bangladesh)
	Relatively stable and fertile volcanic areas (Mt Etna)
Climate	Areas with reliable, evenly distributed rainfall, a lengthy growing season, no temperature extremes (Western Europe)
	Areas with high levels of sunshine (California)
Vegetation	Areas of grassland encourage pastoral farming that supports relatively dense population (Denmark, Pampas of Argentina)
Soils	Areas with deep soils, rich in humus, such as those found in alluvial river basins (the Ganges valley, Paris area, the Nile delta) and which support intensive farming
Resources	Areas with extensive deposits of coal close to the surface (Rhine–Ruhr, Donbas, Yorkshire/Lancashire)
	Areas where intensive farming takes place (Ganges valley, the Low Countries, east coast of China)
Water supplies	Areas with a regular and reliable supply of water. These may be areas with an evenly distributed rainfall (northwest Europe) or heavy seasonal rainfall (the monsoon lands of South East Asia)

Exam tip

Be prepared to make comparative points based on Tables 14 and 15. For example, illustrate the role of resources and water supplies in determining sparsely and densely populated areas.

Key population parameters

Population parameters concern the **distribution** and **density of a population**, as well as the sheer **numbers** involved and rates of change. **Population change** is examined in more detail later (see page 91).

Population distribution is usually displayed by a dot map. Areas that have a large number of people per unit area are densely populated. Areas that have few people per unit area are sparsely populated.

The population density of a country or region is obtained by dividing the total population of that country or region by its total area. This, however, can be misleading as it does not show the variations between densely populated areas and those that are almost unpopulated. Population density is usually displayed by a choropleth, or shading map.

In terms of the **number** of people, in the summer of 2020, the world's population stood at 7.8 billion. Over 6 billion of these lived in developing countries, with 1.3 billion living in developed countries. Half of the world's population live in just six countries: China, India, USA, Indonesia, Brazil and Pakistan. Demographers suggest that by 2043 the global population will reach 9 billion.

Nearly half of the world's population (some 3.4 billion) is under the age of 25, and there are 1.3 billion between the ages of 15 and 24. These are entering their child-rearing years. Their attitudes and responses to birth control will determine whether or not the world reaches 11 billion by 2100. The world's youngest countries (i.e. those with the highest proportions under 15 years) are in Africa.

The key role of development processes

Development processes have been manifested in key moments of history of the planet from the agrarian revolution of Stone Age times to the Industrial Revolution that started in the UK in the eighteenth century. Closer to the present, we have seen the Green Revolution of the 1960s, when new hybrid seeds were introduced, and the technological revolution of the twenty-first century. Development processes have involved the production of more food, industrial development and even the control of many of the threats to population – diseases and pests, for example. Development processes have enabled more people to live on the planet.

It is suggested that a youthful population provides an opportunity for a country to capitalise on its youthfulness and thereby stimulate economic growth – a so-called **demographic dividend** (see page 93). This would be a continuation of the process of development over time, whereby ever-growing numbers of people have made use of the resources available to them to improve their living standards.

Global patterns of population numbers, densities and change rates

References to these have been made earlier, and they are constantly being updated. You are advised to study online data such as the World Population data sheet produced by the Population Reference Bureau (www.prb.org/) to keep up to date with global patterns of population numbers, densities and rates of change.

Various World Population clocks, such as www.census.gov/popclock/, are also useful.

Distribution of a population Its spatial occurrence, i.e. where people are located and where they are not.

Density of population The number of people per unit area, for example per km^2.

Knowledge check 36

What determines the number of people that can live on the planet?

Demographic dividend The benefit a country gets when its working population is much larger than its dependent population (children and the retired).

Knowledge check 37

How can a 'demographic dividend' become a 'demographic debt' as a country develops?

Environment and population

Patterns of food production and consumption

World food production continues to increase, yet the rate at which it is increasing has slowed. Equally, the world has made significant progress in raising food consumption per capita.

Some points about food production and consumption:

■ Much food is produced and consumed locally (especially in poorer countries) or, in the cases of richer countries, it is produced within the same country of consumption.

■ An increasing proportion of food is produced not for domestic markets but for sale in the world's markets. Specialisation and the commercialisation of agriculture drive this change.

■ The actions of transnational corporations (TNCs), and the need for governments of developing countries to raise income through export earnings, have led to the export of many crops from poorer nations, even if there are local food shortages in those countries.

■ Food travels increasingly long distances (measured by food miles), as technological advances in air, sea and freight take place, together with improvements in storage techniques, which ensure food stays fresh.

■ Demand for non-seasonal foodstuffs in richer nations is high (e.g. in winter, strawberries are flown from Chile to the UK).

■ Many countries, e.g. the UK, are net importers of food. This situation also occurs in countries in sub-Saharan Africa, where undernourishment is rife.

■ A few countries are net exporters of food and these countries have few undernourished people (e.g. Canada, Australia, New Zealand and Argentina).

■ Agricultural exports can make up a large percentage of the earnings of poorer countries, but generally form only a very small percentage of the export earnings of developed countries.

There are clear patterns of food consumption. The richest nations consume the most kilocalories per day (between 2,600 and 3,800), including those in North America and Europe, Australia, South Korea, Japan and parts of South America. Large parts of central Africa, Asia and South America consume far fewer calories and here many people can be said to be suffering from under-nutrition – where people consume less than the WHO's recommended daily minimum totals at 1,940 kcal for women and 2,550 kcal for men.

An added problem for those countries where people suffer under-nutrition is that of malnutrition. This is where people may get sufficient calories a day but not have a balanced diet – much of their calorie intake might come just from rice.

Agricultural systems and productivity

Farms can be considered as open systems (see Figure 10). Generally, as an area develops economically, physical factors become less important as human inputs increase in influence.

Making connections

World trade in food forms part of the Global systems topic.

Exam tip

Examiners are likely to provide world maps for this area of study. Make sure you check which show production and which show consumption.

Making connections

Systems theory is a key concept that applies to both physical and human geography.

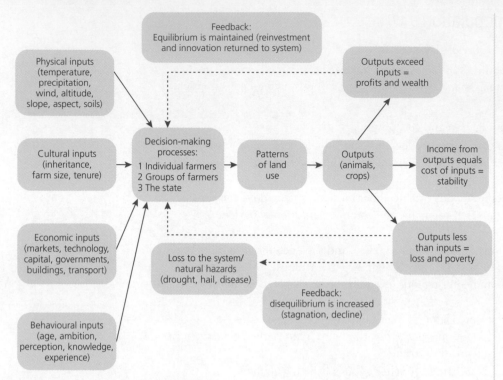

Figure 10 Agriculture as a system

Intensive farming is usually relatively small scale and can either be:

- **capital intensive**, i.e. money is invested in soil improvement, machinery, buildings, pest control and high-quality seeds/animals. There are few people employed and so output is high per hectare and per worker, for example market gardening in the Netherlands, or
- **labour intensive**, i.e. the number of farm workers is high and so there is a high output per hectare but a low output per worker, for example rice cultivation in the Ganges valley of India

Extensive farming is carried out on a large scale over a large area. This varies greatly. There are areas in which, although the labour force is low, there is a high capital input, for example on high-quality seeds/animals or pesticides and insecticides, such as wheat farming in the Canadian Prairies. Other areas still have a low labour force but rely on the sheer amount of land they are farming to provide sufficient output for their needs, for example sheep ranching in Australia.

So, agricultural systems can be a combination of

- **commercial** or **subsistence**
- intensive or extensive
- arable (crops) or livestock (animals) or mixed

Some forms of agriculture are so affected by human influences that they can be regarded as being highly specialised. This can range from high intensities of one crop (monoculture) to highly technological forms such as hydroponics, where soils are deemed to be unnecessary.

As with any system, examine the feedback mechanisms, both positive and negative.

Commercial farming
Where the outputs (crops, animals) are sold.

Subsistence farming
Where the outputs are largely for consumption by the farmer with little or no surplus.

Relationships between climate, soils and agriculture

Despite human factors, such as technological inputs, being increasingly influential in agriculture, the key physical/environmental variables of climate and soils are still important for agriculture on a large scale. In simple terms, some places in the world are more suited for agriculture than others.

You are required to study **two major climatic types** to exemplify relationships between climate and human activities and numbers. You are also required to study the characteristics and distribution of **two key zonal soils** to exemplify the relationship between soils and human activities, especially agriculture. It is easier to combine these in two specific areas of study. One area where climate and soils are ideally suited to agriculture and food production, and hence have an influence on population numbers, are the temperate grasslands of the Canadian prairies.

The Canadian prairies

In the mid-latitude continental interior of North America there are extensive areas of natural grassland. Much of this grassland has, in the last two centuries, been heavily modified by human activity and in particular by farming. People have ploughed up large areas for extensive wheat farming, and they used the rest for equally extensive cattle and sheep ranching. The natural relationship between climate and soils has greatly influenced the type of agriculture being undertaken here. Due to the nature of the agricultural activity, both the numbers and density of people are low.

Climatically, the conditions are as follows:

■ The hottest months have temperatures between 27°C and 21°C, depending on latitude.
■ The coolest months have temperatures between 0°C and –4°C.
■ The annual range, therefore, is approximately 28–22°C.
■ Rainfall totals are usually less than 500 mm, with a summer maximum and some snow in winter.

The soils in this region are known as chernozems, or black earths. They are largely developed on vast expanses of wind-blown (aeolian) silts, known as loess. These silts were laid down on the edges of the huge Pleistocene ice sheets at the end of the ice ages, and were blown there from the northern areas of moraine and outwash plains. These soils are ideal for the natural growth of grassland. Grass growth is vigorous in the spring and early summer, but the dry period in late summer and the frosts of winter slow down the natural processes of decomposition. Consequently, losses of organic material are reduced. The loess parent material also provides a source of calcium carbonate. There is a rich soil fauna, including various burrowing animals, which cause some mixing of the soil profile. Taking these factors together, then, the surface layers of the soil are deep and humus-enriched, fertile and ideal for these forms of agriculture.

There is some debate as to whether the grasslands of the Prairies are entirely a natural response to climate and soils, or whether over time they have developed due to human activity, and are in fact a **plagioclimax**. Botanists have tended to believe that, under the prevailing conditions of climate and soil, the presence of a dense grass sod prevented the invasion and establishment of trees. High evapotranspiration rates were thought to give a competitive advantage to herbaceous plants with shallow, but dense, root systems. Overall, the land, climate and soils are ideal for the farming of grains and meat.

Exam tip

You should undertake a similar study of another area yourself, such as the monsoon areas of India and Bangladesh, or the Mediterranean coast of Europe.

Making connections

This section can be used to illustrate the links between climate, soils and vegetation within the Ecosystems option.

Knowledge check 38

What could be another reason for the lack of trees in the Canadian Prairies?

Plagioclimax Where human interference has permanently arrested and altered the natural vegetation of an area.

Climate change as it affects agriculture

Scientists state that climate change will have significant adverse effects on crop yields, livestock health and tree growth due to higher temperatures, extended heatwaves, flooding, shifting precipitation patterns, and spreading habitats for pests (such as flies and mosquitoes) and diseases (such as wheat and coffee rusts). These will arise from relatively small increases in heat and humidity. Without adaptation, yields of the main cereals in developing countries are expected to be 10% lower by 2050 than they would have been without climate change.

Studies suggest that the primary impact of climate change will be on the poor in tropical countries, mainly through decreased local food supply and higher food prices. The most significant effects are projected for Africa and south Asia, where poverty is highest. Here agriculture accounts for a large share of employment and GDP, and adaptation investment per capita is low. In general terms, the lower the capacity of people to adapt to climate changes, the larger the negative impacts will be.

In 2012, the Food and Agriculture Organisation (FAO) introduced the concept of climate-smart agriculture (CSA). The aim of this was to address both food insecurity and climate change, but the stated outcomes were extremely general:

- Agricultural productivity should be increased in a sustainable fashion.
- Resilience to climate change should be enhanced.
- Agriculture should do its bit to reduce greenhouse gas emissions.

Soil problems

The degradation of soils is the result of human failure to understand and manage them.

Soil erosion

Some activities which cause **soil erosion** are:

- The removal of vegetation by either chopping down trees or overgrazing by animals. In both cases the soil is exposed to the wind and the rain.
- The ploughing of land in the same direction of the slope. This encourages rilling to take place.

Gleying

Gleying is a process of soil formation that takes place when conditions are **waterlogged**, or anaerobic. Under such conditions, the pore spaces are filled with stagnant water, which becomes deoxygenised, causing the reddish-coloured oxidised iron (ferric) deposits in the soil to be chemically reduced to the blue-grey ferrous iron. Reoxygenising causes red-orange patches to appear within the blue-grey soil, this process being known as mottling. Many of the soils of upland Britain show evidence of gleying.

Salinisation

Salinisation is a feature of areas with an arid or semi-arid climate. As moisture is evaporated from the surface, water containing salts is drawn upwards by capillary action. Further evaporation causes the deposition of the salts on the ground surface. The roots of plants that cannot tolerate saline conditions become affected and the plants die.

Making connections

This section has a clear link to climate change (within the Carbon cycle topic).

Knowledge check 39

Give some specific forms of CSA.

Soil erosion The washing away or blowing away of topsoil such that the fertility of the remaining soil is greatly reduced.

Waterlogging Describes the state of a soil when all the pore spaces below a certain depth, known as the water table, are full of water.

Salinisation Occurs when potential evapotranspiration is greater than precipitation and when the water table is near to the ground surface.

Structural deterioration

Soil structure refers to the manner in which individual particles of soil aggregate together. These aggregates are called peds, and they are stuck together by organic matter, and the secretions and mucilages from soil fauna. It is the shape and alignment of the peds that determine the size and number of pore spaces through which water, air and roots can penetrate. Therefore, they influence the agricultural value of the soil.

Examples of **structural deterioration** include:

■ over-cultivation of the soil by growing the same crop in the same field year after year (monoculture)

■ the compaction of the soil by the use of heavy machinery or the trampling effect of animals. This reduces the rate of infiltration into the soil – water flows across its surface and therefore erodes it

Soil management and improvement

Badly soil degraded areas can be managed and improved in a number of ways:

■ Adding fertiliser improves nutrient content. This can be either inorganic (NPK fertilisers which are compounds of nitrogen, phosphorous and potassium) or organic (mainly farmyard manure and crop residues). Organic fertilisers encourage soil organisms and improve nutrient retention. Artificial fertilisers do not do this, and there is concern about their impact on the environment (e.g. eutrophication of lakes and rivers).

■ Planting crops or trees (afforestation) helps to stabilise the soil, and organic matter is returned to the soil through leaf litter.

Various farming practices can also improve the soil, including:

■ crop rotation with fallow periods, allowing soil to replenish nutrients

■ replacing hedgerows to reduce wind erosion

■ improving field drainage to increase aeration

■ ploughing across slopes to prevent gullying

■ liming, providing more nutrients for plant growth and organism development

■ mulching, for example by ploughing in the stubble, increases the organic content and improves nutrient retention

Strategies to ensure food security

The World Food Summit of 1996 defined **food security** as existing 'when all people at all times have access to sufficient, safe, nutritious food to maintain a healthy and active life'. It can also be defined as including both physical and economic access to food that meets people's dietary needs as well as their food preferences.

With the global population expected to grow by 1.2 billion by 2030, pressures on food supplies are set to increase. According to the World Bank, 1.2 billion people still live on less than US$1.25 a day, and more than 800 million people go hungry every day. These people have low levels of food security, or put the other way, high levels of food insecurity. To ensure food security globally, several strategies have been adopted.

Knowledge check 40

Why has salinisation become a major problem in some irrigated areas?

Structural deterioration Occurs when human activity damages the basic structure of a soil.

Exam tip

When considering a form of soil degradation, be aware of the environmental conditions under which the process is operating.

The Green Revolution

This increased crop yields by using new high-yielding varieties (HYVs) that had been developed. Through the use of irrigation, fertilisers, pesticides and mechanisation, high yields per hectare were produced. The Green Revolution has been most successfully applied in Asia, where 90% of all wheat and 67% of rice is produced using HYVs.

However, there are some disadvantages to the Green Revolution, including:

- costly inputs of fertilisers and pesticides, which have led some farmers into debt
- HYVs requiring more weed control
- mechanisation leading to rural unemployment and rural–urban migration

Genetic modification (GM) of crops and animals

Genetic modification entails taking genes (a unit of heredity) from one species and adding them to the DNA (the genetic instructions used in the development and functioning of all known living organisms) of another species. The resultant plant or animal will have some of the characteristics of the donor plant or animal in the resultant offspring.

Advantages of GM:

- Farmers can grow more, because it is easier to fight pests.
- Farmers also use less crop spray (which itself can also be environmentally friendly).

Disadvantages of GM:

- Genes from the GM crop could be transferred to pests. The pests then become resistant to the crop spray.
- Plants can also pollinate weeds, which could then acquire pesticide resistance.

Other approaches to increasing food supply

Cloning: cloning is used in agriculture to produce an identical reproduction of a strong and healthy crop.

Land colonisation: in some developing countries, such as Brazil, government land policies encourage poor farmers to settle on forest lands. Here, each squatter acquires the right to continue using a piece of land by living on a plot of unclaimed public land (no matter how marginal the land) and using it for at least one year and a day.

Land reform: this generally entails the transfer of ownership of land from large (often absent) landowners to smaller resident farmers. Ownership tends to encourage farmers to invest in the land. Examples of land reform can be found in Bolivia, Brazil and India.

Appropriate/intermediate technology solutions: in developing countries, high-tech solutions are not always suitable to the local conditions. Scientists have developed appropriate technology that farmers can learn to use easily, thus becoming self-sufficient.

Knowledge check 41

Give two examples of the use of appropriate technology in agriculture.

Environment, health and wellbeing

Global patterns of health, mortality and morbidity

Mortality

Mortality in children under 5 years of age around the world declined by almost 60% between 1990 and 2018, from an estimated rate of 90 deaths per 1,000 live births to 31 deaths per 1,000 live births. This translates into over 20,000 fewer children dying every day in 2018 than in 1990. The risk of a child dying before their fifth birthday is still highest in the World Health Organization (WHO) African Region (50 per 1,000 live births) – twelve times higher than that in the WHO European Region (4 per 1,000 live births). Nevertheless, nearly 20,000 children worldwide died every day in 2018, and the global speed of decline in mortality rate is disappointingly slow.

Morbidity

Morbidity is measured using the **disability-adjusted life year (DALY)**. This is a measure of overall disease burden, expressed as the number of years lost due to ill-health, disability or early death. Originally developed by the WHO, it extends the concept of potential years of life lost due to premature death to include equivalent years of 'healthy' life lost by virtue of being in a state of either poor health or disability. Most morbidity can be attributed to disease, both infectious and non-communicable.

Infectious diseases include the most serious diseases such as influenza, AIDS, malaria, cholera, yellow fever and most recently coronavirus. Morbidity also arises from non-communicable diseases (NCDs) such as cardiovascular diseases, cancers, chronic respiratory diseases, and diabetes.

The Epidemiological Transition

In 1971, Omran put forward a model relating to population, health and disease – the Epidemiological Transition Model. It states that societies undergo three 'ages' of health.

1 **An age of pestilence and famine:** a period in which mortality is high, with the principle causes of death being infectious diseases and poor maternal conditions, reinforced by nutritional deficiencies.

2 **An age of receding pandemics:** socioeconomic developments and advances in medical science and healthcare mean infectious diseases are reduced and life expectancy increases. Examples of such improvements and advances include better public water supplies and the discovery of penicillin.

3 **An age of degenerative diseases:** as infectious diseases are controlled and people live longer, there is an increased visibility of degenerative diseases (cancers, heart disease). Diseases associated with modernisation and industrialisation (obesity, diabetes) increase.

Some writers have added a fourth stage – an age of delayed degenerative diseases. Here the causes of death are generally the same as the third stage (although dementia is more prevalent), but they occur later in the life cycle as life expectancy increases.

Mortality State of being mortal or susceptible to death. There are various indicators of death, including crude death rate, infant mortality and maternal mortality.

Morbidity Illness. It includes any diseased state, disability or condition of poor health.

Exam tip

The World Population data sheet produced by the Population Reference Bureau (www.prb.org/) enables you to keep up to date with global patterns of mortality.

Exam tip

The WHO provides useful factsheets on NCDs in the world at (www.who.int) – click on the Health topics tab, where NCDs are listed under 'N'.

Omran stated that socioeconomic development is responsible for the movement of a society through these 'ages'. As a model, the Epidemiological Transition Model parallels the Demographic Transition Model (see page 92), from high fertility and mortality rates, with young populations with high levels of infectious disease, to societies with low fertility and mortality rates and ageing populations where NCDs predominate. For the latter to occur, a medical and healthcare revolution has to take place, which tackles and/or controls infectious diseases. The model therefore provides us with a useful reference point for exploring the linkages between health and socioeconomic development.

Exam tip

Make sure you understand the terms 'infectious', 'non-communicable' and 'degenerative' when classifying diseases.

The relationship between environment variables and the incidence of disease

Several writers have sought to examine links between the natural environment and the incidence of disease. For example, there appear to be links between the medical conditions of asthma and hay fever and the prevalence of pollen and dust during warmer times, certainly in the UK.

There also appears to be a direct link with some aspects of the physical environment with the incidence of malaria. Malaria is a tropical disease, associated with a tropical climate. However, it has also been endemic in other parts of the world in the past, for example in Mediterranean Europe. The key is the existence of a particular form of mosquito, which spreads the malaria parasite. Mosquitoes breed in warm areas of stagnant water, and hence are common in flat lowland marshy areas.

In recent years another tropical disease – Ebola – has featured strongly in the news. Is this also a function of the natural environment (another hot environment) or is its spread more complicated? Between 1976 and 2018, the disease was confined to a narrow equatorial climatic belt between southern Sudan and Uganda, west through the Democratic Republic of Congo (DRC) to Gabon, Liberia, Sierra Leone and Guinea. However, some factors that contributed to the Ebola epidemic are deeply rooted in local cultural practices. Bodies are extremely infectious just after death, but local funeral rituals include someone, often a family member, washing and re-dressing bodies prior to burial. People have strongly held beliefs that bad luck, or ill-health, will befall a family that does not carry out funeral rites in a traditional, respectful way. This is not an environmental link.

A further factor is the widespread consumption of bushmeat from the local environment in rural Africa. Although not fully proven, fruit bats are suspected to be a natural wild host of Ebola. Such creatures only exist in the tropical environmental conditions – both climatic and ecological – that allow them to flourish.

Deterministic links between environment and disease are therefore far from clear.

Air quality and health

Air pollution is a major environmental risk to health. By reducing air pollution levels, countries can reduce the burden of disease from stroke, heart disease, lung cancer, and both chronic and acute respiratory diseases, including asthma. The lower the levels of air pollution, the better the cardiovascular and respiratory health of the population will be, both in the long and short term.

According to the WHO, ambient (outdoor air) pollution in both cities and rural areas was estimated to have caused 3.7 million premature deaths worldwide in 2013. Some 88% of those premature deaths occurred in low- and middle-income countries, with the greatest number in the WHO Western Pacific and South East Asia regions.

In addition to outdoor air pollution, indoor smoke is a serious health risk for some 3 billion people who cook and heat their homes with biomass fuels and coal.

Water quality and health

Preventing the spread of water quality-related disease is a major global health challenge. According to the WHO, some of the major challenges include:

- Almost 1 billion people lack access to a safe, clean water supply.
- Two million annual deaths are attributable to unsafe water, sanitation and hygiene, and almost half of these are from diarrhoeal diseases.
- More than 50 countries still report cholera to WHO.
- Schistosomiasis, a disease associated with parasitic worms that live in irrigation ditches and other water courses, has infected an estimated 260 million people.

A biologically transmitted disease: malaria

Prevalence and distribution

An estimated 3.2 billion people (40% of the world's population) are at risk of being infected with malaria. In 2014, an estimated 200 million cases of malaria occurred globally and the disease led to over 590,000 deaths. The burden is heaviest in the WHO African Region, where an estimated 90% of all malaria deaths occur. Children aged below 5 years account for almost 80% of all deaths.

Links to the physical environment

Malaria is caused by parasites of the species plasmodium, which are spread from person to person through the bites of an infected mosquito – the transmission vector. These parasites are spread from one person to another by female mosquitoes of the genus anopheles. Malaria transmission differs in intensity and regularity depending on local factors such as rainfall patterns, proximity of mosquito breeding sites and mosquito species. Some regions have a fairly constant number of cases throughout the year – these are malaria endemic – whereas in other areas there are 'malaria' seasons, usually coinciding with the rainy season.

Malaria occurs in mainly tropical and sub-tropical regions, where the disease is endemic in rainforest and savanna grasslands with at least 1,000 mm of rain per year, and often where the rainfall is seasonal. The parasite needs temperatures of 16–32°C to develop. The incidence of malaria decreases at altitudes over 1,500 m, hence some tropical areas are unaffected, such as the Kenyan Highlands. Like all mosquitoes, the anopheles mosquito breeds well in warm or hot areas of stagnant, standing water, though drainage of these areas has reduced the disease's incidence. For similar reasons, malaria is more common in coastal areas than inland.

Links to the socioeconomic environment

Malaria exacts a heavy burden on the poorest and most vulnerable communities. Within countries where it is endemic, the poorest and most marginalised

Knowledge check 42

Explain how the implementation of good water practices could reduce the global disease burden.

communities are the most severely affected, having the highest risks associated with malaria, and the least access to effective services for prevention, diagnosis and treatment. A number of socioeconomic variables are thought to have influenced a high incidence of malaria:

- **housing quality:** especially densely clustered and overcrowded dwellings
- **unsanitary conditions in the community:** areas with standing dirty water, open waste flows and outlets encourage more mosquitoes
- **occupation:** people working in some jobs are more prone to infection, such as farm workers and irrigation workers
- **levels of education:** researchers have found that, in general, people who have not completed their primary education are more likely to catch malaria

Impact on health and wellbeing

The common first symptoms – fever, headache, and vomiting – appear 10 to 15 days after a person is infected. If not treated promptly with effective medicines, malaria can cause severe illness that is often fatal. Co-morbidities (other illnesses) can make matters worse.

There are further high personal impacts, such as the spending on **insecticide-treated mosquito nets (ITNs)**, doctors' fees, drugs and transport to health facilities. Socially, malaria disrupts schooling and employment through absenteeism, and creates nutrition deficiencies and anaemia in women in malarial regions. As a result, 25% of first babies in some areas have a low birth weight.

In some countries with a very heavy malaria burden, the disease may account for as much as 40% of public health expenditure, 30–50% of inpatient admissions and up to 60% of outpatient visits.

Management and mitigation strategies

Malaria control and ultimately its elimination are inextricably linked with the strengthening of health systems, infrastructure development and poverty reduction. Main interventions include:

- vector controls (which reduce transmission by the mosquito from humans to mosquitoes and then back to humans) achieved using ITNs or **indoor residual spraying (IRS)**
- chemo-prevention (which prevents blood infections in humans)
- case management (which includes diagnosis and treatment of infections)

There has been an expansion in the use of diagnostic testing and the deployment of **ACTs**. This indicates a move away from treating people who have the disease to those who might have it. In 2015, a record number of long-lasting insecticidal nets (LLINs) were delivered to endemic countries in Africa. Another method is that pregnant women receive chemo-preventative treatment during their pregnancy in order to reduce child deaths.

However, emerging drug and insecticide resistance continues to pose a major threat, and if left unaddressed, could trigger an upsurge in the disease. Resistance to artemisinin has been detected in South East Asian countries including Cambodia, Thailand and Vietnam.

Insecticide-treated mosquito net (ITN) Protective nets that are treated with an insecticide, forming a protective barrier around people sleeping under them.

Exam tip

In many examination questions, key words such as physical, social and economic are used. Be clear as to their meaning.

Indoor residual spraying (IRS) The application of insecticides to a person's dwelling, and on walls and other surfaces that serve as mosquito resting places.

ACTs Artemisinin-based combination therapies for the treatment of malaria, combining two active ingredients with different methods of action. They are the most effective anti-malarial treatment available today.

In 2015, US scientists stated that they had bred a genetically modified mosquito that could resist malaria infection. If this laboratory technique works in the field, it could offer a new way of preventing mosquitoes from spreading malaria.

A non-communicable disease: type 2 diabetes

Type 2 diabetes is one of a number of non-communicable diseases (NCDs). There are two types of diabetes:

- **Type 1 diabetes** occurs when the pancreas does not produce insulin. It can occur at any age, but usually before the age of 40. It is usually treated with insulin injections.
- **Type 2 diabetes (T2D)** occurs when the body doesn't produce enough insulin to function properly, or the body's cells don't react to insulin. It is much more common than Type 1 and is often linked with being overweight or obese.

Prevalence and distribution

According to WHO:

- The total number of people in the world with T2D is projected to rise from 382 million in 2013 to 592 million in 2035.
- It is estimated that 80% of the people who have T2D live in developing countries.
- T2D accounted for 11% of global health expenditure in 2013 ($548 billion), a figure expected to be over $600 billion in 2035.
- T2D is associated with roughly 8% of total world mortality, about the same as HIV/AIDS and malaria combined.
- Mortality and disability associated with T2D are particularly high in poor- and middle-income countries where people are unlikely to get the treatment that helps prevent the worst complications of the disease.

Links to the physical environment

Links between most, if not all, NCDs to the physical environment are difficult to establish – this is true for T2D. Some possible links include:

- Increased temperatures within an urban area lead to heat stress, which could add to the pressures already placed on a body due to being overweight or obese.
- Similarly, difficulty of breathing in a polluted environment may add further stress.
- As insulin must be kept in a refrigerator, access to supplies can be challenging in areas with hot climates.

Links to the socioeconomic environment

The WHO states that too much weight gain and obesity are 'driving' the global T2D epidemic. Recent estimates suggest that 1 billion people in the developing world are obese. A number of factors contribute to this:

- **Diet:** high calorie intake is the main factor leading to obesity. In developing countries, rapid economic development has introduced a more 'Western' diet, with less fruit and fewer vegetables, combined with a higher intake of carbohydrates, fatty foods, salt and sugar.
- **Urban lifestyles:** rates of urbanisation in the developing world are considerably higher than elsewhere. A more sedentary lifestyle with a low level of physical activity are features of urban living.

> **Exam tip**
>
> If you have studied another biologically transmitted disease, make sure your notes follow the same sequence of headings.

- **Tobacco use:** smoking is a key risk factor linked to T2D: an estimated 50–60% of adult males in developing countries are regular smokers.
- **Stress:** stress increases blood sugar levels, raises blood pressure and can suppress the digestive process. These are all risk factors in the development of T2D.

Impact on health and wellbeing

Diabetes cannot be cured but can be managed with a mixture of medication and lifestyle change. Longer term, it can lead to heart disease, kidney failure, blindness, and, in extreme cases, amputation of limbs. As most healthcare costs in developing countries must be paid by patients out of their own pockets, the cost of healthcare for T2D creates significant strain on household budgets, particularly for lower-income families.

Management and mitigation strategies

Public awareness of T2D is slowly increasing in many developing countries. In Sri Lanka, for example, awareness is increasing mainly because of local media coverage and the activities of the Diabetes Association of Sri Lanka (DASL). DASL has a walk-in centre in the capital Colombo, where individuals can be screened or take part in structured health programmes at a modest cost. The centre also provides information through workshops, a website and printed materials. DASL also spreads the message that exercise can help reduce the threat of T2D. Exercise areas are being created in many urban centres within the country.

The role of international agencies and NGOs

The World Health Organization (WHO)

The WHO is an agency of the United Nations (UN), established in 1948 and based in Geneva. The work of WHO includes:

- providing a central clearing house for information and research on such features as vaccines, cancer research, nutrition, drug addiction and nuclear radiation hazards
- sponsoring measures for the control of epidemics and endemic diseases by promoting mass campaigns involving vaccination programmes, instruction on the use of antibiotics and insecticides, assistance in providing pure water supplies and sanitation systems and health education for rural populations
- advising on the prevention and treatment of both infectious diseases and NCDs
- working with other UN agencies (such as UNAIDS and UNICEF) and NGOs on international health issues and crises (such as the Ebola crisis in 2014–15)

Non-governmental organisations (NGOs)

A number of NGOs operate in the field of world health. In some developing countries, they often act as alternative healthcare providers to the state, and during times of crisis, e.g. hazardous events, they provide the initial set of responses, certainly in remote areas. They are largely funded by donations from the developed world, mostly by individuals or organisations rather than by nations. One such NGO is Médecins Sans Frontières (MSF).

Making connections

There is a clear connection to Global systems here, where cultural and social globalisation are connected to this area of study.

Exam tip

If you have studied another NCD, make sure your notes follow the same sequence of headings.

Making connections

This section clearly links to the topic of Global governance, where the work of the United Nations is a required area of study.

Knowledge check 43

Describe the work of one identified NGO.

Population change

Factors in natural population change

Fertility

In most parts of the world, **fertility rates** exceed both mortality and migration and are therefore the main determinant of population growth. Several African countries have very high **crude birth rates** of over 40 per 1,000 per year. These include Niger, Burkina Faso, Mali and Burundi. At the other end of the scale, Singapore, South Korea, Japan and Spain have birth rates of less than 12 per 1,000 per year. Fertility varies for a number of reasons:

1   Some countries in sub-Saharan Africa have high birth rates to counter the high rates of infant mortality (some are over 70 per 1,000 live births per year).

2   In many parts of the world, tradition and culture demand high rates of reproduction. One indicator of this is child marriage. Over 60% of women in some countries of sub-Saharan Africa and parts of Asia are married before the age of 18.

3   A key determinant is education, especially female literacy. Knowledge of birth control and more opportunity for employment lower birth rates.

4   Economic factors are important, especially in developing countries where children are seen as an economic asset. In developed countries, this general perception is reversed and the high cost of the child dependency years is often a reason for deferring child rearing.

5   Religion is of major significance, because the Muslim and Roman Catholic religions oppose artificial birth control.

6   There have been instances where politics have influenced fertility. These have been either to increase population (as in Russia and Romania) or to decrease it (as in China with its one-child policy, although this has now been abandoned).

Mortality

Some of the highest **crude death rates** are found in developing countries, particularly in sub-Saharan Africa. The Central African Republic, Lesotho, Chad and Sierra Leone all have death rates of 13 per 1,000 or more. On the other hand, some of the lowest mortality rates are found in countries with a wide range of economic development, for example Andorra, Brunei, Costa Rica and the Maldives. Mortality varies for a number of reasons:

1   Areas with high rates of **infant mortality** have high rates of mortality. Infant mortality is a prime indicator of socioeconomic development and in some areas it is very high, for example Sierra Leone has a rate of 84 per 1,000 live births.

2   Areas with higher levels of medical infrastructure have lower levels of mortality.

3   Poverty, poor nutrition, and a lack of clean water and sanitation (all associated with low levels of economic development) all increase mortality rates.

Natural population change Population change arising from the relationship between crude birth rates and crude death rates. It is usually indicated as a percentage.

Fertility rate The number of live births per 1,000 women aged 15 to 49 per year.

Crude birth rate The number of live births per 1,000 population per year.

Crude death rate The number of deaths per 1,000 population per year.

Infant mortality rate The number of children under the age of 1 year who die per 1,000 live births per year.

Around the world, mortality has fallen steadily due to medical advances. It has been suggested that the world is more willing to control mortality than it is to control fertility.

The Demographic Transition Model (DTM)

The Demographic Transition Model (DTM) is based on the notion that population change over time is closely linked to economic development. It shows how a country's population total will change over time as a result of variations in the birth and death rates. Therefore, it focuses entirely on natural change – migration is not a feature of the DTM. The greater the difference between the birth rate and the death rate, the faster the rate of natural increase and thus the steeper the curve in the growth of total population.

It is generally accepted that there are five stages of the model:
- Stage 1 (High fluctuating): a period of high birth rates and high death rates, both of which fluctuate. Population growth is small.
- Stage 2 (Early expanding): a period of high birth rates but falling death rates. The population begins to expand rapidly.
- Stage 3 (Late expanding): a period of falling birth rates and continuing falling death rates. Population continues to grow but at a slower rate.
- Stage 4 (Low fluctuating): a period of low birth rates and low death rates, both of which fluctuate. Population growth is small.
- Stage 5 (Decline): a later period when death rates slightly exceed birth rates, which causes a population to decline.

Application of the model in contrasting settings

The model directly connects demographic change with social and economic change – there is no attempt to put it into a physical setting. In Stage 1, a poor, agriculturally based, subsistence society is limited in its ability to prevent mortality when food is short or outbreaks of disease occur. A high birth rate will therefore be counteracted by a high death rate.

Stage 2 marks a period of increasing national wealth and as a result there is a significant decline in death rates. In the UK, from 1600, there was an agricultural revolution and by 1750 food supply exceeded population demand. The subsequent Industrial Revolution brought about dreadful working conditions in factories and poverty in rapidly urbanising cities, such as Manchester, but alongside these there were also massive improvements in public health and medical care. So, overall, mortality continued to fall. Today, countries with limited economic development have been able to move into Stage 2, due to the transfer of technology and provision of benefits, such as vaccination programmes, better maternal and neo-natal care and more efficient farming practices.

The model suggests that further increases in wealth cause a fall in birth rate with progression to Stages 3 and 4. Once children are no longer seen as an economic necessity, fewer will be born. However, the model underplays the importance of religious and cultural factors that affect fertility. Birth rates have been slow to fall in countries where there is pressure to have children, for example some Islamic and Catholic countries.

Exam tip

The World Population Data Sheet produced by the Population Reference Bureau (www.prb.org/) enables you to keep up to date with all of these key vital rates.

Exam tip

Most textbooks have a diagram of the DTM. Study it closely.

As stated earlier, the model makes no specific reference to the impact of migration on population change. The fall in death rates attributed to economic growth in Western Europe was also helped by emigration – large numbers of people left Italy and Ireland, for example, for the New World in the mid-nineteenth and early twentieth centuries, and this relieved some pressure on land resources. The progress of some developing countries through the demographic transition is currently slow or even static. For example, Botswana, Lesotho and Swaziland, which have a high incidence of HIV/AIDS, have maintained high death rates despite signs of economic progress.

Knowledge check 44

How might a country regress through the Demographic Transition Model (DTM)?

The demographic dividend

The concept of the demographic dividend was introduced earlier in this book (page 78). A country's fertility rate falls during its demographic transition as its level of development increases. The result is fewer dependent children and relatively more productive teenagers and young adults in the population. A large body of young adult people with high aspirations can boost economic growth, provided there is investment in education and employment, and little emigration. Reasons for this include:

- A large young adult workforce serves as a powerful magnet for 'footloose' global companies to invest in the country.
- Workers with fewer children invest some of their income, contributing to financial growth.
- Women become more likely to enter the formal workforce, promoting greater gender equality.
- Salaried workers quickly become consumers, so global retailers and media corporations view these countries as important emerging markets.

However, a demographic dividend is not always delivered when **population structure** changes. Due to a lack of social development, some countries fail to make the most of their human resources. A large working-age population is a wasted opportunity if levels of education, especially numeracy and literacy, are weak. Good governance is essential.

Population structure
Refers to the make-up of the population of a country. The most studied form of structure is that of age and sex composition and is represented by a population pyramid.

Knowledge check 45

Explain how changing lifestyles may affect the demographic dividend.

Age–sex composition

Standard population pyramids can be drawn to represent each stage of the DTM and the structures of individual countries are often compared with these pyramids. In simple terms, the stages are as follows:

- Stage 1: has a triangular shape which shows a wide base (indicating a high birth rate) and sloping sides (indicating a high death rate) forming a peak with few elderly people.
- Stage 2: the sides become slightly more steep, showing a decrease in the death rate.
- Stage 3: the sides are steep but the base is less wide, showing a decline in the birth rate.
- Stage 4: a dome-shape – the sides are steep all the way up to the 65 age group and there is an increase in the height of the pyramid, showing more older people.
- Stage 5: the base narrows further, showing the recent lower birth rate.

Exam tip

Most textbooks have diagrams of population pyramids associated with each stage of the DTM. Study them closely.

Population structure is closely linked to **dependency**. Hence, some writers refer to the following categories.

- **A youthful population**, which has a large number of younger people in proportion to the working population. This puts pressure on education, housing and maternal health services.
- **An ageing population**, when there are a large number of older people in proportion to the working population. This puts pressure on the working population to work longer, and the age of attaining a pension may have to rise to provide the increasing numbers of older people with pensions and old-age care.

Cultural controls

A number of cultural controls on population change have been referred to in previous sections:

- **Attitudes to women:** women are discriminated against in many cultures, either openly or implicitly. Features such as low levels of education, early marriage (even child marriage) and forced/arranged marriages are common in some areas.
- **Gender preferences:** also feature in many societies, for example, the desire for a male heir. This preference can be loosely based on economic factors, for example the need for a male to till the land. However, it can be based on societal norms – families may continue to procreate till they have a son, and in some extreme cases, female infanticide has taken place.
- **Religion:** this influences attitudes to artificial contraception and abortion. In some African, Arabian and North American societies polygamy is prevalent.

International migration

When examining population change, the balance between immigration and emigration (**net migration**) must also be considered. **Migration** can be local or within a country, in which case it will not change the overall population total for any country. **International migration** does change the total population of a country. It is estimated that there are over 270 million international migrants in the world.

Economic migrants form by far the greatest proportion of international migrants. It is thought that poverty drives the typical economic migrant. Examples of movements of economic migrants include the following.

- In the late nineteenth and early twentieth century, transatlantic migrations involved the movement of large numbers of people from Europe to North America. Migrants moved from countries such as Ireland, Italy and Norway to the new industrialising economy of the USA, where 'fortunes could be made'.
- Since 1980, there have been substantial increases in migration to (or within) Western Europe, firstly from countries such as Greece, Spain and Portugal, then from North and West Africa, and most recently from central and Eastern Europe.
- Since 2015, thousands of migrants have fled Syria, Afghanistan, South Sudan and Libya to seek refuge in European countries. Germany, the richest country within the EU, has taken in the largest number.

Much smaller proportions of people are forced to move as **refugees** and **asylum seekers**, although the numbers are still large. The **UNHCR** estimated that there were over 26 million refugees in 2019. Refugees are unable (or unwilling) to return to

The dependency ratio
The relationship between the economically active (working) population and the non-economically active (dependent) population.

Making connections
International migration is an important feature of both Global systems (e.g. shrinking world) and Changing places (e.g. changing population characteristics).

Net migration The difference between the numbers of in-migrants and out-migrants.

Migration This involves a permanent or semi-permanent change of residence.

International migration The movement of people across national frontiers, for a minimum of 1 year.

Economic migrant A person who moves voluntarily for work or to improve his/her social conditions.

Refugee A person who, owing to a fear of being persecuted for reasons of race, religion, nationality, membership of a particular social group, or political opinion, has fled his or her country and sought protection in another country.

their country of origin for fear of persecution. Historically, countries affected by civil war, by persecution of minority groups on grounds of religion or ethnicity, or governed by political regimes that punish dissent, have all produced large refugee populations. A refugee is a successful asylum seeker. There has been a worldwide growth in applications for asylum. The UNHCR reported that 4.2 million people submitted asylum applications in 2019. It is then up to the government of the receiving country to decide if they qualify as refugees and can be allowed to stay. In 2018, the highest numbers of asylum applications were submitted in Germany and the USA, mostly people from Syria and Afghanistan.

Causes of international migration

Various factors are said to influence international migration:

- **Push factors:** the negative aspects of the current place of residence. They include factors such as lack of employment, low wages, poor housing, poor educational opportunities, political persecution, natural hazards, starvation and war.
- **Pull factors:** the attractions of the place of destination. Often they are the converse of the push factors: better employment and educational opportunities, better housing and social services, higher wages, family integration and political stability.
- **Perception:** the subjective view that a person has of an environment, derived from personal experience, the experience of others and from the media. If the perceived push or pull factors are strong enough to overcome the forces of inertia (cost of moving, disruption of social networks) migration will occur. It is often this perception that is the basis of the decision-making process.

It is no surprise that migration should be growing in an increasingly interconnected world. The combination of technological change, improved transport infrastructure and economic growth has made mobility easier and more desirable to many.

Implications of international migration

Table 16 Demographic, environmental, social, economic, health and political implications of international migration

Factor	Implication for country of origin	Implication for country of destination
Demographic	Reduction in birth rate	Increase in birth rate
	Imbalanced population structure – ageing population	Imbalanced population structure-youthful population
Environmental	Abandonment of land and houses	Pressure on land for new housing stock
Social	Reduced pressure on social services (health/education)	Increased pressure on social services (health/education)
	Loss of skills ('brain drain')	Multiculturalism benefits/issues
Economic	Reduced levels of unemployment	Skills shortages and employment vacancies filled
	Remittances sent home	Cheap labour source
	New skills introduced by returnees	Tax benefits to the state
		Demographic dividend created (see page 93)
Health	Reduced pressure on health facilities	Increase in incidence and variety of infectious diseases
		Increased pressure on health services
		The issue of 'health tourism'
Political	Issues regarding declining regions – investment needed?	Arguments to reduce immigration
		Rise of right-wing political parties

Asylum seeker A person who has applied for refugee status and is waiting for a decision as to whether or not they qualify.

UNHCR The United Nations High Commissioner for Refugees.

Knowledge check 46

Where are the major sources of refugees in the world today?

Exam tip

When analysing or evaluating any of these implications, try to support the points you make with references to specific issues and/or locations.

Principles of population ecology and their application to human populations

Population growth dynamics

The key parameters of population growth dynamics are birth and death rates. As we have seen, the number of births in a population is limited not only by biological factors (e.g. the ages of both the male and especially the female of the species) but also by cultural factors, such as age of marriage and attitudes to contraception. Also, it was noted that one important factor influencing birth rates was a specific death rate: infant mortality. If offspring are more likely to die young, then birth rates increase to compensate.

Such parameters are also seen in the natural world, and hence some writers have linked population growth to ecology, i.e. the natural world. Furthermore, as in the natural world, human populations are affected by uncontrollable events which may impact on them – for example, natural hazards and disease (e.g. the coronavirus pandemic). Changes in food supply have also impacted human populations in the past, and still do have an effect in developing areas of the world, such as the famines in East Africa in the 1980s. Some scientists have translated this into ecological statements:

- When biotic potential (the reproductive capacity) is greater than environmental resistance (limiting factors such as famine and disease), then there is population growth.
- Similarly, if limiting factors outweigh biotic potential, then populations will decline.

Overpopulation, underpopulation and optimum population

It is estimated that the world's population will reach 9 billion before 2050. Some have questioned whether the world can cope with ever-increasing numbers of people. Are there enough resources to go around?

When studying the balance between population and resources, three concepts are considered:

- **Overpopulation** occurs where the population is too large for the resources available. This relationship also depends on the level of technology available to help to make good use of resources, as well as on the climate and physical limitations of the area. Overpopulation can cause unemployment and outwards migration.
- **Underpopulation** occurs where there are not enough people living in an area or country to utilise the resources efficiently. An increase in the number of people would therefore result in a higher standard of living. Underpopulated areas tend to have inward migration and low unemployment.
- **Optimum population** is stated to exist where the resources available can be developed efficiently in order to satisfy the needs of the current population and provide the highest standard of living. However, as technology develops, the optimum population will increase. This theoretical concept has not been achieved anywhere in the world.

> **Exam tip**
>
> One way of illustrating these concepts is to draw a simple line graph with GDP per capita on the y-axis, and total population on the x-axis. Try to draw it.

Implications of population size and structure for the balance between population and resources

These can be examined in the context of a country – **Japan** (population 127 million).

Population decline has started to take place in Japan and will continue into the future. The cause is purely natural – death rates (11 per 1,000) exceed birth rates (8 per 1,000). Experts attribute Japan's population decline to the high cost of raising children in the country, the growing number of women who choose to work longer and have a career rather than have children, and Japan's reluctance to accept immigrants.

Whether or not such population decline is a problem that *needs* a solution is open to debate. Population decline has many benefits, but in Japan it will also present significant problems. Rural areas tend to experience significant outflows of young people who move to urban areas, in addition to low birth rates. Such depopulation is driven by young people being drawn to the life, education, and employment opportunities of urban areas. A consequence of this depopulation is the increasing amount of abandoned houses, and their associated land, left unoccupied when the last resident dies.

The increasingly inverted structure of Japan's population pyramid, with fewer young people than old people, means that it will be very difficult to generate the tax revenues necessary to pay for the healthcare needs of the elderly. Japan's older population (over 65) is currently around 28% of the total. In 2050, this proportion is expected to be 36%. As the elderly population grows, the financial burden of healthcare in Japan will become substantial, and there could very well be a shortage of labour in the healthcare industry.

It is possible that Japan's population will drop to just 96 million by 2050. While birth incentives and immigration incentives are suggested as the solution to bring the young workers necessary to support the country's ageing population, there is a national reluctance for either of these to take place. Slowing population growth and an ageing population are shrinking its pool of taxable citizens, causing the social welfare costs to spiral upwards. This has led to Japan becoming the most indebted industrial nation in the world.

The concepts of carrying capacity and ecological footprint

Development, the living standards of people and in turn the consumption patterns of a population influence **carrying capacity**. The latter is also a function of attitudes. As a society becomes more 'westernised', consumption rates increase. This can be evidenced by the increasing consumer consumption rates of people in emerging economies, such as China and India, for designer goods.

Using assessment of the **ecological footprint**, it is possible to estimate how much of the Earth (or how many planet Earths) it would take to support humanity if everybody lived a given lifestyle (and consumption rate). You can calculate your own personal ecological footprint using the website http://ecologicalfootprint.com/.

Some have developed these ideas by suggesting that each year should have an 'Earth Overshoot Day' – the day when the productive capacity of the planet has been used up for that calendar year. In 2000 it was 1 November, while in 2020 it was 22 August.

> **Exam tip**
>
> You could analyse another country in order to examine the links between population size/structure and resources. Good examples would be a Gulf state (high immigration), or a sub-Saharan country (young population).

> The **carrying capacity** of an area refers to the largest population that the resources of a given environment can support.

> **Ecological footprint**
> Refers to a measurement of the area of land and/or water required to provide a person (or society) with the energy, food and other resources they consume, and render the waste they produce harmless.

The Population, Resources and Pollution model

Human activities can affect both the biotic (natural) and abiotic (physical) conditions of an environment. As human culture has developed over time from hunting and gathering to agriculture and then into industrial societies, the impact on these environmental conditions has grown to a level that many believe is unsustainable. The **Population, Resources and Pollution (PRP)** model (Figure 11) illustrates several important relationships between people and their environment, and offers a 'big picture' view of human–environment interactions.

The PRP model shows that humans, like all other organisms, acquire resources from the environment. The acquisition of resources, for example through coal mining, alters both the biotic and abiotic environment. Surface coal mines cause deforestation and disrupt habitats. They can also lead to soil erosion that pollutes nearby streams. Hence there is a link in the model between resource acquisition and pollution.

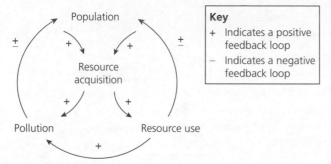

Figure 11 The Population, Resources and Pollution model

Resources extracted from the ground are then put to use. For example, after being mined, coal is burnt in power stations. Other minerals extracted from the ground are also fashioned into finished products, such as motor vehicles and machinery. Such activities can produce pollution, indicated in the model by the link between resource use and pollution.

The model also contains examples of **negative** and **positive feedback loops**. Positive feedback loops tend to reinforce trends, whereas negative feedback loops have a balancing effect. Some of the positive feedback loops shown on the PRP model are of concern to environmentalists. For example, continued soil erosion from the world's farmlands resulting from poor agricultural practices such as overgrazing (often due to population pressures) could cause a marked decline in food production, which could affect the global population.

On the other hand, negative feedback is shown when resource acquisition and use, especially that permitted by technology, can promote population growth. The efficient harvest of food has made it possible for larger numbers of people to inhabit the Earth.

Contrasting perspectives on population growth

The case for pessimism: Thomas Malthus

Thomas Malthus stated in his 1798 work *An Essay on the Principle of Population* that population growth was geometric (1, 2, 4, 8, 16 etc.) while food supply could only grow arithmetically (1, 2, 3, 4, 5). This, inevitably, leads to a point where carrying capacity is exceeded, leading to a food shortage. Malthus believed that increasing

Negative feedback loops Act in such a way to nullify any change that has taken place. They are the main means of controlling biological systems and regulating equilibrium.

Positive feedback loops Act such that one change leads to a further change, which in turn stimulates the first one in a repetitive cycle. In this way change is perpetuated.

Making connections

Systems theory (including feedback mechanisms) is a key concept that applies to both physical and human geography.

the food supply dramatically was not possible, and so food shortages would lead to a series of population checks, which would reduce the population size. These checks would either increase the death rate (a 'positive check') or reduce the birth rate (a 'preventative check'). Examples of positive checks include war, famine and epidemic. Preventative checks occur when individuals realise that there may not be enough food to support a family, and so they opt for later marriages or sexual abstinence (note, this was before artificial contraception was widely available).

Malthus believed a series of cycles would occur, in which carrying capacity was exceeded, each leading to a period of population check. Malthus argued pessimistically that preventative checks would never reduce the population sufficiently and that we would never escape from cruel positive checks – starvation, war and disease are inevitable.

In 1972, 'The Limits to Growth' report, produced by the **neo-Malthusian Club of Rome,** suggested that there would be a sudden decline in global population in 100 years due to the overuse of resources. This decrease in population would return the world to a state of balance with the resources on the land.

The case for optimism: Ester Boserup

Ester Boserup, a UN agricultural economist, argued in 1965 that it was the size of a population that determined the level of its food supply. By examining the development of agriculture in a number of regions, Boserup concluded that as population size reached carrying capacity, societies were forced to make radical agricultural changes to ensure there was enough food. For example, she noted that the people of Java in Indonesia had over time adapted their farming practices several times to feed a larger population. They had moved from long fallow farming (land is farmed for 1–3 years and then left to recover for 10–20 years) to short fallow, and then to annual cropping (a crop is harvested annually from a piece of land) to multiple cropping (a piece of land produces more than one annual crop).

Such optimists do not regard population size as a problem, because humans can invent solutions to any food shortages. They argue that agricultural advances over the last century, such as the Green Revolution, and more recently, the development of genetically modified foods, show this. In essence, 'necessity is the mother of invention'.

Other writers have continued this optimistic theme. **Julian Simon** summed up the optimistic perspective in his book *The Ultimate Resource* (1981) by suggesting that the ultimate resource is human ingenuity.

Global population futures

Health impacts of global environmental change

Ozone depletion

The **ozone layer** shields the Earth from most of the harmful ultraviolet radiation from the sun. It is this radiation which causes an increase in the incidence of skin cancer and eye cataracts. Ozone depletion occurs when the rate at which the ozone layer is formed is less than the rate at which it is destroyed. In recent years, this depletion has manifested itself by the emergence of a 'hole' in the ozone layer over

Knowledge check 47

Give an example of when positive feedback in the Population, Resources and Pollution (PRP) model can create environmental problems.

Exam tip

You could research the views of other writers on this controversial topic such as Stephen Emmott, Bjørn Lomborg and Danny Dorling.

Ozone layer A concentration of the gas ozone located in the stratosphere at an altitude of between 10 km and 50 km above sea level.

Antarctica and Australia/New Zealand, and the Arctic. The damage is believed to have been caused by the use of chlorofluorocarbons (CFCs) in refrigeration and insulation, which break down ozone, although the EU and other countries have now banned their use*.

(* Note: the hydrofluorocarbons (HFCs) that are used now are super greenhouse gases.)

Skin cancer is a disease associated with exposure to the sun, and made worse by more UV radiation reaching the surface through ozone 'holes'. Rates of malignant melanoma vary around the world, with Australia being the worst affected. Within the UK, the incidence of skin cancer has increased within the over-55 year age group, as its rise has been linked to the package holiday boom to southern Europe that dates from the 1960s.

Cataracts are a form of eye damage that can eventually lead to blindness if not treated. As with skin cancer, increasing rates of cataracts have been linked to UV radiation. Incidence rates are greater in tropical areas, at higher altitudes where UV levels are higher, and also where the UV radiation can be reflected into the eye from surfaces such as snow.

Climate change

The WHO has estimated that up to a quarter of a million people per year will die as a result of climate change. Climate change is likely to have impacts on health in several ways:

- direct impacts from extreme weather events such as more frequent and severe storms and floods, and heatwaves
- indirect impacts from environmental and ecological change such as:
  - **Thermal stress:** hot and humid weather can cause illness, such as dehydration, heat exhaustion and death. Heatwaves also exacerbate other health risks such as smog, wildfires and vermin infestations. Older people and those with chronic medical conditions are most vulnerable. In 2020, *The Lancet* stated that there had been a 50% increase in heat-related deaths of over 65s since 2000, especially in the developed world.
  - Emergent and changing distributions of **vector borne diseases**: it is predicted that the prevalence of existing diseases such as malaria and dengue fever will increase as the mosquito vectors spread further around the world following warmer and wetter weather. Areas most at risk include China, Mexico, Turkey and the southern USA. Further, it is thought that other emergent diseases could have a greater incidence such as the Zika virus in Latin America, the West Nile virus in North America and Lyme disease within Europe (including the UK).
  - **Agricultural productivity:** higher latitude areas will see an increase in crop yields, in the variety of crops (e.g. maize and soya beans) that can be grown and in the length of the growing season. However, in some lower latitudes, there will be more incidence of thermal stress for plants and livestock, and yields may fall.
  - **Nutritional standards:** it is suggested that there will be a decline in red meat consumption in the developed world which will benefit the environment through less livestock production and less forest clearance. However, crop failure in areas most at risk from the damaging effects of climate change will have a negative impact on the nutrition of the people living there.

> **Exam tip**
>
> When analysing or evaluating any of these impacts, try to support the points you make with references to specific issues and/or locations.

Prospects for global population

The world's population will continue to grow more slowly than in the recent past. In 2000, world population grew by 1.24% per year. In 2015, it grew by 1.18% per year – still an additional 83 million people annually. It is projected to increase by more than one billion people within the next 15 years, reaching 8.5 billion in 2030, and to increase further to 9.7 billion in 2050 and 11.2 billion by 2100 (Table 17).

Table 17 Present and projected population totals by area (2015–2100)

Area of world	2015 (millions)	2050 (millions)	2100 (millions)
Africa	1,186	2,478	4,387
Asia	4,393	5,267	4,889
Europe	738	707	646
Latin America and Caribbean	634	784	721
North America	358	433	500
Oceania	39	57	71
World	7,349	9,725	11,213

Source: UN Population Division (2015)

Projected distributions

More than half of global population growth between now and 2050 is expected to occur in Africa. Africa has the highest rate of population growth, growing at 2.6% per year. Consequently, of the additional 2.4 billion people projected to be added to the global population between 2015 and 2050, 1.3 billion will be in Africa.

Asia is projected to be the second-largest contributor to future global population growth, adding 0.9 billion people between 2015 and 2050, followed by North America, Latin America and the Caribbean and Oceania. Europe is projected to have a smaller population in 2050 than in 2015 (Table 17).

Critical appraisal of future population–environment relationships

The developed world

The populations of most European countries will decline in the next generation, and in the cases of Germany and Russia (and also Japan), the decline will be dramatic. The contraction of the population will leave a relatively small number of workers supporting a very large group of retirees, particularly as **life expectancy** increases. In addition, it may be left to the smaller, younger generation to pay off the national debts the older generation incurred.

One solution to this problem is immigration. However, a major issue is that some westernised countries have cultural problems integrating immigrants. This was evidenced during and after the recent EU referendum campaign in the UK, and although Europeans have tried to integrate immigrants – particularly with migrants from the Islamic world – they have found it a challenge. While Japan does not have a history of integrating migrants, the USA has historical sources of immigration, particularly from Mexico. However, some right-wing politicians within the USA, such as the former President Trump, want to restrict migration from Mexico.

The developing world

When Western countries went through their demographic transition, both mortality and fertility fell gradually over a period of a century or more. Since the

Knowledge check 48

How will the populations of China, India and Nigeria change in the future?

Life expectancy At birth, the average age to which a person is expected to live.

1960s, demographers have been surprised to see how fast the equivalent transition happened in Asia and Latin America. In countries like China and Brazil, it took just three or four decades for the fertility level to plunge from more than six to less than two. The experience in Asia and Latin America led demographers to expect a similarly rapid transition in Africa. However, this has not been the case. Over the past two decades, it has become clear that fertility is falling much more slowly in some countries in sub-Saharan Africa, such as Nigeria, than it did on other continents. According to the UN, of the people added to the planet in this century, one in five will be Nigerian.

There is, however, another variable that needs to be considered when examining future population growth: education. Educating girls in particular has been found to be one of the best ways of bringing down fertility in the long term. However, progress in female education is slow, and governments in Africa and elsewhere need to make access to family planning more widely available. Education is the key to global population and environmental futures.

Exam tip

This is a very topical area of study. Try to keep up to date with it by using online resources such as Gapminder.

Case studies

You are required to **study one country/society experiencing specific patterns of population change** to illustrate and analyse the character, scale and patterns of change that have taken place. You should include relevant environmental and socioeconomic factors, and analyse the implications for that country/society. Examples could include the ageing population of Japan, or the migration of Eastern Europeans into the UK.

You are also required to study a **specified local area** to illustrate and analyse the relationship between place and health related to its environment, socioeconomic character and the experience and attitudes of its population. Examples could include the work of health charities in specific villages in the developing world, or field study and/or online research based on communities within the UK.

Summary

After studying this topic, you should be able to:
- understand the environmental contexts for population, both physical (e.g. climate, soils) and human (e.g. development, causes of change) and their impact on population numbers and densities
- know and understand the global patterns of food supply, the agricultural systems that produce it, and the factors that cause variations in it, such as climate and soil types
- explain how climate change and soil problems can impact on food supply and food security
- know and understand the patterns of world health, and the economic, social and environmental factors that cause variations in it
- know and understand the causes and impacts of two different diseases, and the strategies to manage and mitigate them – one biologically transmitted and one non-communicable
- explain how population changes over time, and assess the impact of contrasting physical and human settings on population change
- analyse the reasons for, and consequences of, international migration
- understand the links between population dynamics and ecology, and evaluate the various theories associated with the link between population and resources
- examine the prospects for future population growth within a changing climate, with particular regard to health matters and population–environment relationships

Resource security

Resource development

The concept of a resource

The traditional approach to a **resource** considers that it is the act of exploitation which converts a commodity into a resource, i.e. it emphasises the use of the commodity. More radical approaches stress the exchange value of a resource. On commodity markets, profits can be made without doing anything to a resource, owners buy at one price and sell on at another without perhaps ever seeing the resource or even taking ownership of it in the sense of storing it. Such speculative behaviour, which also applies to the purchase of land in anticipation of its future potential value, views resources in a different way.

Classification of resources

Resources may be classed as being either **renewable** (or **flow**), or **non-renewable** (or **stock**). In terms of renewable resources, distinction should be made between those that occur constantly, independent of human activity, such as solar radiation, winds and tides, and those which can be maintained, depleted or sometimes increased by humans, such as soil and natural vegetation, such as trees.

Renewable (flow) Renewable resources are those which can be replenished with the passage of time and should therefore always be available for humans to use.

Non-renewable (stock) Non-renewable resources are those that have been created by long-term physical processes over geological timescales and therefore cannot be replaced.

Non-renewable resources include:
- fossil fuels such as coal and oil
- rocks such as limestone and granite
- precious and semi-precious stones such as diamonds and emeralds
- metallic elements contained within rocks such as iron ore, copper and rare earth elements

Stock resource evaluation

Key determinants of the value of stock resources are the quantity and quality of the materials in their natural state. High levels of both of these should make them economically viable to extract. However, the evaluation of their worth is more complex than this, and a McKelvey box can be used to illustrate this (Figure 12).

Resources includes all the deposits of a mineral/commodity – those that are discovered, undiscovered or unviable. **Reserves** are those parts of the resources that can be economically, technically or legally extracted. Of course, this level can change as commodity prices may fluctuate, and technology is always developing. More resources can become reserves over time.

Resource A resource is any feature of the environment which can be used to meet human needs. Some nations have an abundance of resources (e.g. water, oil and minerals), whereas others have less.

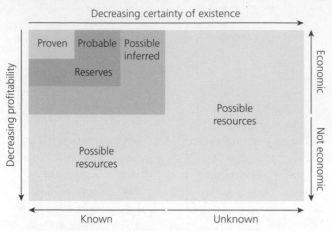

Figure 12 A McKelvey box of Resources v Reserves

Those resources for which geology is a key factor, e.g. mineral resources, can also be subdivided into four further categories:

■ **Proven (measured) reserves:** where the quantity and quality of a resource are well-known enough that they can be viewed as being economically viable.
■ **Probable (indicated) reserves:** where the quantity and quality of a resource are well-known enough that further evaluation of their economic viability can occur.
■ **Possible (inferred) reserves:** occur where the quantity and quality can only be estimated on the basis of limited evidence. There may be issues regarding access.
■ **Possible resources:** are based on broad geological knowledge of the existence of other situations similar to those where resources have already been discovered. Hence these resources are largely undiscovered, and confidence of their existence is low.

Natural resource development

There is a range of influences on the rate at which resources are explored, exploited and developed:

■ **Population 'need':** as population numbers grow it is inevitable that demand for resources will rise, particularly for food, water, energy and building materials.
■ **Population 'wants':** even when basic needs are met, demand for resources continues to grow as people aspire to ever higher standards of living.
■ **Economic development:** natural resources are commodities that are important in their own right or form the raw materials for manufactured goods. This drives continued exploitation in order for economic growth to occur.
■ **Exploration and innovation:** some resources may be unused until it becomes possible and/or economic to exploit them. Some resources may not be needed until new products and technologies that make use of them are developed.

The concept of the resource frontier

A resource frontier is a location where a resource is brought into production for the first time. Such areas are typically geographically remote (peripheral) and lack infrastructure (transport, social and economic), which often has to be brought in by

> **Exam tip**
>
> This is a very conceptual area of study. In an exam, you are likely to be asked to differentiate between each of these terms – be clear as to their meaning.

the developers of the resource. Once exploitation of the resource begins, the area starts to develop with the arrival of associated activities.

The concept of resource peak

Knowledge check 49

Identify some resource frontiers for one extraction industry you have studied.

This is perhaps best illustrated by the notion of 'peak oil'. Peak oil refers to the point in time when crude oil production reaches its maximum level (Figure 13). After peak production, supply will decline and oil prices will rise.

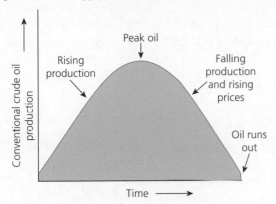

Figure 13 The concept of peak oil

Crude oil and natural gas are fossil fuels, with a finite supply that will eventually run out. The peak is more important than the point in time at which they actually run out because of the economic factor of rising prices as supply falls.

There is considerable debate over when global peak oil production will occur. This is because of the price of oil. There are two ways of defining how much oil is left in the ground:

- **oil resources:** the total quantity of hydrocarbons that are underground, including those that may not be economically feasible to extract and those that are still undiscovered
- **oil reserves:** the amount of oil that it is economically viable to extract

The price of oil determines the economic feasibility of oil extraction, and therefore, as the price of oil changes, the volume of reserves changes. Oil that was too expensive to extract in certain market conditions may well be viable in other market conditions. The volume of oil reserves is therefore always changing because of changes in price, and the volume of oil resources is also changing as new discoveries are taking place, partly as a result of exploration and partly as better extraction methods are developed.

Exam tip

Keep up to date with this area of study. For example, the coronavirus pandemic of 2020 and 2021 impacted the oil industry significantly – loss of demand will have impacted the concept of peak oil.

In addition, the steep price rises expected after peak oil could be limited by developing alternative supplies – such as that from tar sands and oil shales by the process known as fracking. However, although these are likely to increase the reserves of oil, they are unlikely to decrease dramatically the price of oil.

Sustainable resource development

There are several dimensions to this issue – economic, social and environmental. The economies of the developing world should be enabled to attain the same levels of growth as the developed world. People in the developing world are entitled to have access to their fair share of the world's resources. The environmental impacts of resource development should not endanger the same environments where the exploitation is taking place.

However, achieving **sustainability** is not straightforward. Two approaches could be taken:

- **supply management:** increasing the supply of resources by, for example, increasing exploration, increasing research, or developing alternatives
- **demand management:** reducing the consumption of resources by, for example, changing lifestyles, increasing efficiency, recycling, reducing population or having greater governance over the exploitation and use of resources

A key aspect of sustainable resource development is minimising the environmental impact of resource development and use.

Environmental Impact Assessment (EIA)

Environmental Impact Assessment (EIA) aims to predict environmental impacts at an early stage in project planning and design, find ways and means to reduce adverse impacts, shape projects to suit the local environment and present the predictions and a range of options available to decision-makers. An EIA involves the following stages:

- **screening:** to determine which projects or developments require a full or partial impact assessment study
- **scoping:** to identify which potential impacts are relevant to assess, and to identify alternative solutions that avoid, mitigate or compensate adverse impacts on biodiversity. This can include the options of either not proceeding with the development, finding alternative designs or sites which avoid the impacts, incorporating safeguards in the design of the project, or providing compensation for adverse impacts
- **assessment and evaluation of impacts and development of alternatives:** to predict the likely environmental impacts of a proposed project or development, including the detailed consideration of alternatives
- **decision-making** on whether to approve the project or not, and under what conditions
- **monitoring, compliance, enforcement and environmental auditing:** to monitor whether the predicted impacts and proposed mitigation measures occur and to ensure that unpredicted impacts or failed mitigation measures are identified and addressed

Natural resource issues

Energy: global patterns of production, consumption and trade

Note: the following sections refer to the period of time before the coronavirus pandemic of 2020 and 2021, which had a major impact on global patterns of energy.

Energy in total

The **production** of energy around the world varies greatly. In the developed world, we tend to think of energy being derived from sources such as fossil fuels (coal, oil, natural gas), nuclear power and renewable sources. However, we must remember that

Sustainability The degree to which meeting the needs of today does not compromise the ability of future generations to meet their own needs.

Environmental Impact Assessment (EIA) A process of evaluating the likely environmental impacts of a proposed project or development, taking into account socioeconomic, cultural and human health impacts, both beneficial and adverse, prior to decision-making.

Knowledge check 50

Illustrate the carrying out of an EIA in relation to a resource development project.

for a large proportion of the world (possibly over 2 billion people), energy is derived from sources such as fuelwood and biomass. The following sections will concentrate on the 'developed world' sources. During the first part of the twenty-first century the production of energy has increased for all fuels except coal.

The emerging economies (**BRICS**) have accounted for most of the net growth in energy **consumption** over the past decade. Growth in Chinese consumption has largely driven this increase and has been responsible for the world's largest incremental rise in primary energy consumption each year since 2000. On the other hand, energy consumption by **OECD** countries has experienced a steady decline. For example, energy consumption in the EU in 2017 fell to its lowest level since 2005, mainly due to the decline of manufacturing.

Oil

Oil remains the world's leading fuel, with over 30% of global energy consumption in 2018, although it has lost market share in recent years. Since 2015, growth in global oil **production** has frequently been greater than that of global oil **consumption**. Oil reserves are mainly in the middle east (especially Saudi Arabia – 25% of the world's reserves), Russia and the former Russian states, e.g. Kazakhstan, with the USA and Canada producing a large amount from their tar sands deposits. Other major producers include Venezuela, Nigeria and Angola. The USA is largely responsible for the growth in production in the world. It accounted for the majority of the non-**OPEC** output and in doing so took over from Saudi Arabia as the world's largest oil producer.

The largest consumption per capita is in North America, Western Europe, Australia, Saudi Arabia and Russia. However, countries outside the OECD now account for the majority (51%) of actual global oil consumption and they account for all of the net growth in global consumption.

In terms of global oil **trade**, the major importers are China, Western Europe and the other emerging economies, with declines in the USA and Japan. 61% of global oil consumption is traded between countries – this equates to huge movements of oil around the world in tankers and through pipelines.

Natural gas

Since 2015, global natural gas **production** has grown by almost 2%. The USA remains the world's leading producer and has had the largest increase, accounting for 75% of global growth. Gas reserves are greatest in Iran, Russia, Qatar and Turkmenistan.

Globally, natural gas accounts for approximately 25% of primary energy **consumption**. Global natural gas consumption continues to grow, with growth mainly in the OECD countries and the emerging economies. However, consumption has fallen in the EU, especially in Germany, Italy, France and the UK. The US, China and Iran have had large increases in consumption.

Global natural gas **trade** has declined in recent years, driven by falls in pipeline exports from Russia. The UK, Germany and Ukraine all reduced their pipeline imports markedly. Interestingly, global trade in LNG (Liquefied Natural Gas) has grown (33% of all gas trade in 2018). There have been increased imports into the UK and China. Qatar remains the largest LNG exporter (32% of global exports).

BRICS An acronym for Brazil, Russia, India, China and South Africa.

OECD A group of over 30 nations that includes many of the world's most advanced countries (USA, UK, Germany etc.) but also emerging countries such as Mexico, Chile and Turkey. It does not include the BRICS nations.

Knowledge check 51

Which two countries produce the most energy?

OPEC Organization of the Petroleum Exporting Countries: a cartel which sets output quotas in order to control prices of crude oil. There are currently 13 members of OPEC, mainly countries in the middle east, South America, Africa and Asia.

Coal

Global coal **production** continues to fall, with large declines in both China and Ukraine. Major sources of coal include China and India, the USA, Australia and Indonesia. **Trade** is largely from these latter countries to China and India, which, as well as producing coal are also the largest consumers. Coal **consumption** is static, with its share of global primary energy consumption remaining at 28%. Consumption in the EU continues to fall.

Nuclear and hydro-electric

Global **nuclear** output is growing, but at a slow rate. Nuclear power accounts for about 5% of global primary energy consumption. Increases in South Korea, Switzerland and France are partly offset by declines in Japan, Belgium and China.

Global **hydro-electric** output growth is also slow, led by growth in China and India. Hydro-electric output accounted for 7% of global primary energy consumption in 2018.

Renewables

Consumption of other renewable energy sources – in power generation as well as transport – continues to increase, reaching about 4% of global energy consumption. Renewable energy used in electricity power generation has also grown and accounts for 26% of global generation (2019). China has recorded the largest incremental growth in renewables, followed by the USA, while growth in Europe's leading players – Germany, Spain and Italy – has reduced. Globally, wind energy accounts for more than half of renewable power generation growth and solar power generation has grown even more rapidly, but from a smaller base.

Knowledge check 52

Which countries produce the most energy from biofuels?

Exam tip

This is a very topical area of study. Up-to-date data can be found at www.iea.org/

Ore minerals: global patterns of production, consumption and trade

The term 'ore mineral' covers a range of commodities which do not share the same characteristics. In terms of production, consumption and trade, ore minerals can be classified as those that are:

- **abundant:** iron, aluminium and silicon
- **scarce:** chromium, nickel, copper, lead, tin, tantalum and zinc
- **precious:** gold, silver, platinum

Production and consumption

Each of these minerals has its own areas of production (Table 18), though they may have similar areas of consumption, which generate different directions of trade. In general terms, areas of consumption are the developed nations of the world (North America, Western Europe, Australia) and some emerging nations (China, India and Russia). Some of these have their own areas of production and hence are not involved in trade – for example, Australia and Canada are large producers of mineral ores. On the other hand, several countries in central and southern Africa have extensive deposits of mineral ores and depend heavily on exporting them. For many years, these countries exported their minerals in a relatively raw state, with the purification being largely undertaken in developed countries. Over recent decades, such refining, **beneficiation** and smelting activities have moved back to the producing countries – sometimes for economic cost-savings, and sometimes to raise financial returns for the host countries.

Beneficiation
The concentration/improvement of raw materials, usually metallic ores, close to where they are extracted.

Table 18 Top producers of mineral ores

Mineral ore	Producing nations
Iron ore	China, Australia, Brazil, India
Bauxite	Australia, China, Brazil, Guinea
Copper	Chile, China, Peru, USA
Tin	China, Indonesia, Peru, Bolivia
Chromium	South Africa, India, Kazakhstan, Turkey
Lead	China, Australia, USA
Zinc	Australia, China, USA, Peru

Consumption of all the major mineral ores has increased in the last 50 years. Perhaps the most dramatic has been the rise in the use of aluminium. Others, such as iron ore and copper, have also shown increases, largely driven by the economic growth of China and India.

By 2018, China was the major importer of iron ore, with two-thirds of total global imports, though consumption here has fallen recently, causing a global surplus of steel (the main product) to occur. It also imported almost half of the world's copper. The main consumers of aluminium are the USA and European nations, mostly for use in the aeronautics industry.

Trade

The economic dominance of the Asian Pacific (China, South Korea and Taiwan) is a key factor influencing world trade patterns. Although there has been a recent slowdown in economic growth in the Asian Pacific region, the demand for ore minerals remains much higher than elsewhere in the world, so the trade pattern is likely to stay the same. Other major importing areas are Western Europe and Japan.

Water: global patterns of availability and demand

Sources

Over 97% of all water on Earth is salt water (saline). Of the remaining water (freshwater), about two-thirds is stored as ice sheet and glacier, mostly in Antarctica and Greenland. Most of the remainder is stored underground as groundwater and only 0.006% of all freshwater is stored in rivers. About two-thirds of the world's population relies on groundwater aquifers. While freshwater is a renewable resource, pollution results in diminishing supplies of clean, safe water.

In theory there is enough freshwater on the planet for the people who live on it – the issue is one of distribution. There are areas of water surplus and areas of water scarcity.

Areas of **water surplus** are those where:
- the climate offers regular, plentiful and reliable rainfall
- there is effective governance of the available water supply
- there are efficient storage and distribution networks and/or low demand such as in areas of low population

Geographically, they tend to be in the equatorial regions and temperate areas of the world.

Knowledge check 53

Explain the role of TNCs in the price of mineral ores.

Making connections

This section is clearly connected to the Water cycle, where the distribution of the world's water stores is examined in more detail.

Water scarcity

Water scarcity is a relative concept and can occur at any level of supply or demand. It can be a social phenomenon (i.e. a product of affluence, expectations and behaviour) or the physical consequence of poor or altered supply patterns – for example, resulting from climate change.

According to the UN, around 1.2 billion people (one-fifth of the world's population) live in areas of water scarcity, and this figure is expected to rise to 1.8 billion by 2025. In addition, water scarcity in some arid and semi-arid places is expected to displace between 25 million and 700 million people. Another 1.6 billion people (one-quarter of the world's population) face water shortage where countries lack the necessary infrastructure to take water from rivers and aquifers.

Water scarcity can be a result of two mechanisms: physical water scarcity and economic water scarcity. Physical water scarcity is a result of inadequate natural water resources to supply a region's demand. Economic water scarcity is a result of poor management of the sufficient available water resources (Table 19).

Water scarcity The point at which the combined demand of all water users (e.g. agriculture, industry) cannot be satisfied fully.

Exam tip

Research online maps of areas of water surplus and water scarcity.

Table 19 Global variations in water scarcity

Type of water scarcity	Locations	Commentary
Physical water scarcity	Southwest USA, northern Mexico, North Africa, middle east, central Asia, southern India, southeast Australia	Located in desert and semi-arid areas with low or very seasonal rainfall Water supplies insufficient for the demand from large populations in some areas Poor management of water sources
Approaching physical water scarcity	The near and middle east, southern Africa, parts of Spain and Turkey, southwest USA – Texas, central Mexico, northeast Brazil	Likely to be related to climate change altering rainfall patterns and raising aridity Increased pressure on water sources due to rising populations and economic growth
Economic water scarcity	Large parts of sub-Saharan Africa, central America, northern India, parts of South East Asia	Ample supply but a lack of financial resources to make safe water available to the populations Closely related to poverty
No water scarcity	Northern and eastern USA, Canada, Europe, northern and southern China, much of South America	Well-managed water resources in areas with high and non-seasonal rainfall

Demand components

Irrigation provides the main demand for water, accounting for almost 70% of available freshwater. Some 22% of global water is used by industries, particularly those involved with smelting metals, paper making and electricity production. In low- and middle-income countries, 82% of water is used in agriculture compared with just 30% in high-income countries.

Countries of the developed world have the highest levels of domestic demand, with the USA and Australia being the largest per capita users. Affluent areas of the world use water for personal purposes (domestic appliances, sanitation and leisure use) because they have the wealth to do so. Sub-Saharan Africa has the lowest domestic per capita usage.

Knowledge check 54

Identify some environmental impacts of increased demand for water.

Geopolitics

The geopolitics of energy

Energy supplies are distributed unevenly, and are often a long way from the consumers. Some energy supplies are in politically unstable areas, or are transported through areas where there is conflict. There is a risk of disruption to energy supply (as happened to Ukraine when the Russian firm Gazprom threatened to raise prices excessively in 2014), and of environmental damage (such as the nuclear explosion at Chernobyl, USSR, in 1986). There is also concern regarding some unstable countries developing nuclear power which could then lead to the development of nuclear weapons. India and Pakistan have nuclear power stations but have not signed the Nuclear Non-Proliferation Treaty, and there are worries, mainly from the USA, about Iran building new nuclear power stations.

Other possible geopolitical issues involving oil and gas exploration include the potential exploitation of the Arctic and Antarctica. The seabed of the former will become available for exploration and possible production with the continued retreat of the ice shelf in the Arctic Ocean due to climate change. As a precursor to possible conflict here, Russia has, mischievously to some, placed a Russian flag on the seabed of the Arctic Ocean. The possibility of oil exploration on the continental shelf around the seas of the Falkland Islands has brought further disagreement between the UK and Argentina.

The various recent conflicts within the middle east – Kuwait, Iraq, Libya – have all been linked to the vast oil reserves in those countries. Such conflicts are ongoing, with the activities of the terrorist groups Islamic State (Daesh) and Al Qaeda being of increasing concern in that part of the world.

It is clear that there are links between energy and global politics. The G8 group of nations have met several times to try to agree a global energy strategy. So far they have agreed to work together to increase stability of energy markets, improve investment, diversify energy mix, ensure security of energy infrastructure, reduce energy poverty, and work on sustainable development of energy to address climate change issues.

At a national and local scale, energy can also have a significant political impact. The issue of 'fracking' may continue to feature in discussions in the UK in the coming years, such as in the Fylde (Lancashire) and North Yorkshire areas.

The geopolitics of water

Water also has a geopolitical role. Competition over limited resources of water between neighbouring countries can result in water resources being used as a weapon in a stressed political situation. For example, water supply is being used as a lever in the political crisis between Israel and the Palestinians of the Gaza Strip (see page 118). It is also causing tensions in northeast Africa where Ethiopia has built a new dam on the Blue Nile – the Grand Ethiopian Renaissance Dam (GERD) – that may restrict water flow through Sudan and Egypt (see page 114).

Two factors will almost certainly make political issues regarding the world's water supply of even greater importance in the future:
- Population growth is set to continue, each individual demanding their share of a limited resource.
- Climate change is likely to create hotter drier conditions in tropical zones, areas already experiencing limited water availability.

Exam tip

You could research the development of new oil and gas resources in the Caspian Sea, and the associated risk of conflict between Russia, Azerbaijan, Kazakhstan, Georgia and Turkey.

Exam tip

You could research the development of new oil and gas resources in the eastern Mediterranean Sea, and the associated risk of conflict between Turkey, Greece, Cyprus, Lebanon and Israel.

Making connections

This section is connected to Global governance – the Global Commons.

Knowledge check 55

Provide details of one area where a water resource has become a 'political hotspot'.

The geopolitics of ore minerals

Ore minerals have a number of geopolitical issues including aspects of trade, interdependency, access to rare (but needed) minerals, wider environmental consequences, the role of TNCs and the dominance of China in the demand for ores. Several of these can be examined in the context of **Rare Earth Elements (REEs)**.

Although REEs occur in many locations, they are usually found in minute quantities and are difficult to extract – this is why they are known as 'rare'. Global production is over 200,000 tonnes per annum. China contains over 50% of the world's reserves of REEs and supplies over 70% of current global production.

REEs have become increasingly important. They have a huge range of uses, for example in smartphones, tablets and flat screen TVs. They are also vital to 'green' technology such as wind turbines, solar panels, hybrid vehicles, LED lighting and rechargeable batteries. They are used in satellite communications, medical imaging in hospitals and in fibre-optic cables.

REEs are at the centre of a twenty-first-century power struggle. China, which effectively controls global production of REEs, has been accused of **resource nationalism** by its critics, by threatening to potentially 'starve' the EU, USA and Japan of their supplies. These countries have expressed concerns to the WTO that Chinese export restrictions are an attempt to conserve supplies for its own high-tech, green and military industries. China's dominance in this field has caused new areas of exploration to appear in Australia, Canada, South Africa, Thailand and Brazil. It is an aspect of current economic geography that will be interesting to monitor as demand for REEs is unlikely to fall.

> **Rare Earth Elements (REEs)** A group of chemically similar, metallic elements largely contained within the lanthanide group of the periodic table (atomic numbers 51–71).

> **Exam tip**
>
> The geopolitics of all resource production and consumption is a highly topical area of study. Try to stay up to date with this important topic.

Water security

Sources of water and components of demand

Water can be sourced from:
- rivers and reservoirs – surface supplies
- aquifers – underground supplies
- seawater which has had the salt removed by desalination

The three main components of demand are:
- agriculture – mostly through irrigation
- industry – used in generating power, as a coolant in many factories, and as a raw material in a wide range of manufacturing processes such as textiles, paper and food processing
- domestic use – washing, sanitation and consumption

Water stress

Water stress refers to the difficulty of obtaining sources of freshwater for use during a year, and may result in further depletion and deterioration of available water resources. In 2018, the World Resources Institute (WRI) found that over 35 countries face extremely high levels of water stress. This means that more than 80% of the water available to agricultural, domestic, and industrial users is withdrawn annually, leaving businesses, farms and communities vulnerable to scarcity.

> **Water stress** The ratio of total annual water withdrawals to total available annual supply.

> **Exam tip**
>
> Research online maps of areas of water stress.

It is estimated that 1.1 billion people in the world are water-stressed, meaning that they do not have access to sufficient potable water. Most of these are living in developing countries. According to the Falkenmark Water Stress Indicator, a country is said to experience 'water stress' when annual water supplies drop below $1700\,m^3$ per person per year. Water stress is intensifying in countries such as China and India, and in the middle east.

Extremely high levels of water stress do not necessarily mean that a country will fall victim to water scarcity. Countries facing extremely high water stress can implement management and conservation strategies to secure their water supplies. Singapore, for example, has a high potential for water stress. The country is densely populated and has no natural freshwater lakes or aquifers, and its demand for water far exceeds its naturally occurring supply.

Singapore has invested heavily in technology, international agreements and responsible management, allowing it to meet its freshwater needs. Advanced rainwater capture systems currently contribute 20% of Singapore's water supply, 40% is imported from Malaysia, **greywater** reuse adds 30%, and desalination produces the remaining 10% of the supply to meet the country's total demand. There are plans to reduce all trans-boundary imports of water by 2060, with a consequent increase in the use of recycled and grey water, and desalination.

The relationship between water supply and physical factors

Climate

In general terms, precipitation determines the amount of water that becomes available for consumption. Its seasonality also determines when water is more or less available to all users. Temperature determines the degree and amount of evapotranspiration that takes place to further influence water amounts.

Another climatic factor concerns the fact that the amount of available freshwater is decreasing because of climate change. Climate change has caused glaciers to recede, lower levels of stream and river flow, and shrinking lakes.

Many aquifers have been over pumped and are not recharging quickly enough from diminishing natural flows of water. Furthermore, although the total freshwater supply may not be used up, much has become polluted, salted or unsuitable for drinking or for use by industry and agriculture. Therefore to avoid a global water crisis, farmers will have to strive to increase productivity to meet growing demands for food, while industry and urban areas in particular find ways to use water more efficiently.

Geology and drainage

Geology influences the amount of water that can seep into groundwater stores to become available through aquifers, and drainage affects the amount of water flowing through rivers from which people can draw water.

Consider two different parts of the world: Botswana and central Wales. Botswana is a country in southern Africa that has severe water shortages. Due to the high temperatures, over 90% of the rain that falls is evaporated away. As a result, most rivers are highly seasonal and run dry in the dry season. Due to the lack of rainfall, the water table is often deep, and drops further in the dry season. Conversely, in central Wales,

Greywater Recycled water from showers, baths and washing machines. It is *not* water that has come into contact with faeces, either from the toilet or from washing nappies.

Making connections

This section is connected to the Water budget (the Water cycle), where the flows between different stores of water are examined.

the relief is high, temperatures are much cooler and rainfall totals are high, at 1800 mm a year. The relief of the land makes it ideal to store water on the impermeable granitic rocks, with there being deep narrow valleys such as the Elan Valley. Due to the altitude of the area, there is sufficient 'head' to transfer the water from storage reservoirs by gravity to urban areas such as Birmingham, over 100 km away to the east.

> **Exam tip**
>
> Be prepared to evaluate the relative importance of physical factors in the water supply in an area you have studied.

Strategies to increase water supply

Table 20 Strategies to increase water supply

Strategy	Commentary
Catchment dams and reservoirs	Involves surface water storage behind dams and in reservoirs or the raising of the existing levels of dams to increase their storage capacity. Requires substantial capital investment, and could also have a negative impact on the environment and local communities – people would have to be displaced
Desalination	Involves utilising seawater by building new desalination plants; alternatively, existing desalination plants could be expanded to increase their capacity Involves a significant capital outlay, together with substantial energy costs, and would produce wastewater that would require safe disposal
Aquifer recharge	Involves the collection of rainwater followed by an artificial recharge of an aquifer with the collected water
Diversion and inter-basin water transfers	Involves the interlinking of river basins or their rivers to transfer river water from surplus basins to other basins by pumping
Gravity transfers	Involves the interlinking of water management areas to transfer water resources from surplus basins to other basins by gravity
Rainwater harvesting	Involves the collection of rainwater on rooftops for domestic use
Direct seawater use	Involves the direct use of seawater without removing the salt. This could be used mostly for industrial cooling and municipal purposes (but not for drinking or irrigation)

Environmental impacts of a major water supply scheme: the Grand Ethiopian Renaissance Dam

Introduction

In 2011, Ethiopia announced plans for a huge new hydro-electric (HEP) dam – the Grand Ethiopian Renaissance Dam (GERD) – to be built across the Blue Nile River close to the country's border with Sudan. The scheme was estimated to cost $4 billion. Since then, Egypt and Ethiopia have repeatedly clashed over the dam and its future operations. Ethiopia says GERD will help pull many people in the country out of poverty and boost development, but Egypt believes it will endanger its share of the crucial Nile river waters.

> **Exam tip**
>
> Study one detailed example of a water transfer project, such as the South–North Project in China.

Ethiopia's view

Ethiopia, with a fast-growing population of 100 million, sees the dam, which will be Africa's largest when completed, as a key element of its future economic development. It aims to become Africa's biggest power exporter with a projected capacity of more than 6,000 megawatts.

> **Exam tip**
>
> Study the area of the GERD in an atlas, or an online source, e.g. Google Earth.

Egypt's view

Egypt fears losing reliable access to the water of the Nile, as it believes the dam will reduce the flow of the whole river after the Blue Nile's confluence with the White Nile at Khartoum in Sudan. Egypt similarly has a fast-growing population of over 100 million and is highly reliant on the Nile for 90% of its freshwater. The country is also highly dependent on the river for its agriculture and its industry.

Disagreement

After initial agreements had been reached, extended negotiations regarding annual water release amounts, and strategies to deal with river flows during drought years, have struggled to proceed. In particular, Egypt wants a minimum annual release of 40 billion cubic metres (bcm) of water after completion of the dam. Ethiopia wants a minimum annual release of 35 bcm of water. By November 2020, negotiations were still continuing despite the dam being largely complete.

Climate change

Water is a limited resource, and this is why these two powerful North African nations have been locked in this dispute. Furthermore, water security in the region is also threatened by climate change. Africa's 1.2 billion people stand to suffer the most from climate change, while contributing to it the least. According to the International Energy Agency (IEA), energy-related CO_2 emissions in Africa represented only 2% of cumulative global emissions (2019).

Parts of North Africa are warming at a faster pace than elsewhere, and climate experts have said warming Indian Ocean waters have contributed to more powerful cyclones and severe locust outbreaks in the region. Cyclone Gati (November 2020) struck nearby Somalia and was the most powerful cyclone ever to affect the area.

African heads of state are increasingly concerned about such events in the future, and want the rest of the world, including the top polluters China and the United States, to step up and contribute to Africa's efforts to adapt. Africa has 15% of the world's population, yet according to the **African Union**, it is likely to shoulder nearly 50% of the estimated global climate change adaptation costs.

Strategies to manage water consumption

Table 21 Management of water consumption – agricultural demand

Strategy	Commentary
Agricultural rainwater harvesting with **fertigation**	Boost the productivity of currently rain-fed crops by applying fertilised water during dry spells. It requires the construction of small storage reservoirs
Canal lining and piped water conveyance	Line irrigation canals with cement/plastic to reduce seepage; use of pipe systems to transport water and reduce water evaporation
Channel and irrigation control	Introduce more active controls to limit spill losses through automated measurement of flows, together with better scheduling of irrigation flows
Genetic crop development	Development and adoption of varieties that enable farmers to attain higher yields with less water; includes genetic engineering
Mulching	Cover soil with protective plastics to prevent water evaporation
Soil techniques/ no-till agriculture	Techniques to reduce tillage; laser land levelling to reduce runoff and drain land better, and conserve water
Sprinkler irrigation	Increases yields and irrigation efficiency (e.g. through reduced evaporation) compared with the use of open irrigation channels
Sprinkler conversion to micro-sprayer	Use micro-sprayers in areas where drippers are not practical; consumes less water than standard sprinklers
Retaining stubble on the land	The keeping of stubble (rather than burning it) improves soil water retention and increases moisture levels

Making connections

This section is connected to the Carbon cycle, where responses (mitigation and adaptation) to the impacts of climate change are examined.

African Union (AU)
The AU is a pan-African organisation whose goal is to facilitate the continent towards peace and prosperity. Its aims are to boost development, eradicate poverty and bring Africa into the global economy.

Fertigation The injection of fertilisers and other water-soluble nutrients into an irrigation system.

Table 22 Management of water consumption – industrial demand

Strategy	Commentary
Better housekeeping	Better water management within factories with increased monitoring
	Reduction of leaks in water pipes in industrial facilities
	Reduction of water consumption in washing of industrial facilities and equipment
Pulp and paper industries	Use less water during process of bark stripping from logs. After treatment, reuse of water evaporated in the pulp formation process; during the pulping process, use concealed units to avoid water loss through spray and evaporation
Power industries	Replace traditional water-cooling system with an air-cooling system
Food and beverage industries	New technology replacing water lubrication with Teflon/silicon-based products, thereby eliminating need for water
Mining industries	Water use to suppress dust on haulage roads can be reduced significantly by adding a chemical additive that aids in dust suppression
	Pump out water sitting in unused mines, and reuse it in operations
Water pressure reduction	Reduce pressure in industrial systems, thereby reducing losses from leaks

Table 23 Management of water consumption – domestic/municipal demand

Strategy	Commentary
Management of leaks	Reduction of leaks on commercial and public premises, and in household connections and pipes; reduction of water loss through leak detection and repair in distribution networks
Laundry machines and dishwashers	Use of water-efficient machines/dishwashers
Dual-flush toilets	Installation of water saving dual-flush toilets
Showerheads and taps	Installation of water-efficient showerheads/taps with aerators and pressure controllers to keep the water flow at desired levels
Household landscaping	Introduction of water-efficient techniques (e.g. mulching) in private gardens
Wastewater reuse	Reuse of treated municipal and industrial wastewater as industrial cooling water

Exam tip

There are a lot of strategies in these tables – be prepared to evaluate/assess their relative impact/success.

Sustainability issues

Water stewardship is at the heart of water sustainability. This is concerned with achieving a greater balance between supply and demand. It can be achieved in a number of ways.

Virtual water trade

One way to save water is to not use it in the first place, especially in those countries where the production of food uses disproportionately large amounts of water. The solution is to import that food rather than try to grow it. In this way, the water used in the production process is being used in the area of production and 'traded' by

Water stewardship
The socially equitable, environmentally sustainable and economically beneficial use of water.

the importing country. The amount of water used in the production process can be estimated to give the product a 'virtual water value'. This concept is an extension of the idea of comparative advantage, but is based on water efficiency.

Conservation

Many large food and drinks companies are pursuing programmes to reduce their water consumption. Kellogg's, for example, has introduced a range of water-saving initiatives at some of its manufacturing plants and has reported achieving savings of up to 70%.

For many companies, the biggest challenge in reducing water consumption lies within their often complex supply chains. For a company such as Nestlé, the scale of this challenge is enormous, as the company works directly with over 700,000 farmers. However, Nestlé explicitly recognises that engaging with its geographically widespread supply chain is critical if the company is to meet its water stewardship goals.

Recycling

Programmes to reduce water use are often linked to wastewater treatment and recycling. Smithfield Foods, one of North America's food processing companies, has undertaken a project to help conserve aquifers in North Carolina. Here the company's slaughterhouse at Tar Heel, which opened in 1992, initially withdrew 9 million litres of water each day from two local aquifers. The company installed a 'water rescue system' designed to recycle over 4.5 million litres of water per day which in turn allowed the company to increase production while reducing water demands.

Greywater

While greywater may look 'dirty', it is a safe and beneficial source of irrigation water in a garden or allotment. If released into rivers, lakes or estuaries, the nutrients in greywater become pollutants, but to plants, they are valuable fertiliser.

The easiest way to use greywater is to pipe it directly outside and use it (almost immediately) to water ornamental plants or fruit trees. Greywater can be used directly on vegetables as long as it doesn't touch edible parts of the plants. In any greywater system, it is essential to put nothing toxic into the system – such as bleach, dye, detergents, bath salts and cleansers – though bio-degradable soaps can be used.

Groundwater management

Aquifers can be recharged by human activity. This can be achieved by pumping water underground, diverting rivers on to areas underlain by permeable rocks, and diverting storm water into recharge wells – a technique used extensively in parts of Australia.

Water conflicts

The importance of water means that it can be used for political purposes. As stated earlier (see The geopolitics of water on page 111), conflict hotspots over water resources are common at a variety of scales: international, national and local. The case study of the GERD in Ethiopia (page 114) illustrates a situation where water is a key element in a conflict zone at an international scale. The following example illustrates the issue at a regional, national and local scale.

Making connections

This section is connected to Global systems – specifically the activities of transnational corporations (TNCs).

Knowledge check 56

Name one city where water is recycled on a large scale and briefly state how.

Gaza and Israel

Background

Gaza is a strip of land 40 km long and 10 km wide, bounded by the Mediterranean Sea, Israel and Egypt, and is home to 2 million people. Originally occupied by Egypt, and after a 38-year period following the 1967 Six-Day War, when the territory was run by Israel, Gaza is now ruled by the militant Islamist group Hamas.

Israel imposes a blockade on the territory, restricting the movement of goods, including fuel, and people. Furthermore, Egypt has restricted movement across Gaza's southern border. Gaza is a densely populated area, consisting of urban areas, refugee camps and desolate land. Its population density is one of the highest in the world, at over 5,800 people per square kilometre.

Water resources

Gaza has a warm temperate climate with mild winters and hot, dry summers. All Mediterranean climates rely on winter rainfall for water supplies, but Gaza's pattern seems to be a two-year cycle, with one wetter winter, followed by a drier one. Gaza therefore has little rain and no major freshwater source to replenish its diminishing underground water supplies. Only 6% of the piped water meets WHO quality standards. Treatment of wastewater and sewage is another headache. Gaza relies on wastewater treatment plants that are either working beyond their capacity or were constructed as temporary installations. As a result, about 90 million litres of untreated or partially treated sewage is pumped into the Mediterranean sea every day.

Water scarcity

While under the Israeli administration, there were no plans for an effective water management strategy, and a structural scarcity of water supply existed. This situation continues. Gazan communities have high birth rates, and the rate of natural increase is one of the highest in the world, at 6.0% per annum. Water availability per capita therefore continues to decrease, and soon it is possible that there will not be enough for drinking water, let alone for other purposes. Access to water supplies is limited for the population, being available on average between 6 and 8 hours in every 24 on only a few days a week.

The Gaza aquifer is currently being over-pumped in order to try to meet demand. Israel, which also has its water supply problems, has been tapping into the aquifer from outside Gaza and taking some water which would naturally have replenished it. Overuse causes the water table to fall, and the deficiency is filled naturally by water seeping in from nearby sources, in this case saline aquifers and from the sea itself. Seawater has been detected 1.5 km inland. The whole of the Gaza aquifer is therefore threatened with salinisation. In some places, irrigation water is already so saline it has damaged crop yields, such as citrus fruit.

Extensive amounts of the water infrastructure have been damaged or destroyed by various Israeli military actions in the twenty-first century. Up to 12 wells have been rendered inoperable and more than 6,000 roof storage tanks damaged. Essential materials, such as cement, pipes and pumps, can no longer be accessed – Gaza is essentially cut off from the world, as Israel and Egypt control its borders. Illegal tunnels exist but these are inadequate to bring in larger pipes and equipment.

Exam tip

Study the area of Gaza in an atlas or an online source, e.g. Google Earth.

Resolution

There is ongoing debate about who is responsible for the deteriorating water situation. The pressure group Amnesty International has reported that the water situation in Gaza has reached a 'crisis point', with 90–95% of the water supply in the territory contaminated and unfit for human consumption. In return, the Israeli Water Authority (IWA) claims this report was 'biased and incorrect'. The IWA claims that Israel has met its obligations under the Oslo Accords (1995), which laid down the amount of water that Israel is obliged to provide for the Gazans.

Gazans have managed to make some improvements through ingenuity. A new, rather primitive, wastewater plant has been built at Rafah. Building materials came from the remains of the wall which used to divide Gaza from the Sinai Peninsula and which had been blown up. New shallow wells have been dug which should supply water without placing too much pressure on the aquifer. Desalination plants are, however, needed to improve the quality of aquifer water. UNEP (United Nations Environment Programme) has estimated that an investment of over $1.5 billion is needed to restore the aquifer.

Energy security

Sources of energy

Introduction

Primary energy sources are those used for power in their natural or unprocessed form. They include wood, fossil fuels (oil, natural gas and coal) and power from uranium (nuclear), as well as renewable sources such as solar or tidal. Some of these may be burnt directly – such as wood, coal, oil and gas – or they may be manufactured or converted into **secondary sources** – such as electricity for commercial, industrial and domestic use, and petrol and diesel for transport.

Generally, more developed countries derive their energy from oil, natural gas, coal, hydro-electric power and nuclear resources. Developing countries increasingly use the same fuels, but also use wood, peat and animal waste, especially in rural communities.

Energy mixes

Table 24 **Energy mixes** of a selection of countries (2018) (%)

Country	Coal	Oil	Natural gas	Nuclear	Renewables
USA	29	2	32	20	17
UK	4	1	40	20	35
Switzerland	0	0	1	38	61
France	2	1	8	71	18
New Zealand	5	0	12	0	83
Iceland	0	0	0	0	100
Japan	32	5	33	5	25
Germany	38	1	13	10	38
Sweden	1	0	0	40	59
Australia	60	2	20	0	18
Brazil	5	3	12	4	76
China	66	1	3	5	25

Source: IEA

Exam tip

Up-to-date data on energy mixes can be obtained from online sources such as the IEA.

Energy mix The proportion of different sources of energy used by households and industry together with that used in electricity generation in a country.

The relationship between energy supply and physical geography

Climate, geology and drainage can each play a role in energy supply. This section relates the main sources of energy to one or more of these physical geography aspects.

Coal

Coal is a combustible black or brownish-black sedimentary rock occurring usually in layers, or strata, or veins called coal beds or coal seams. Australia has an abundant supply of coal from such geological formations – approximately 9% of the world's reserves. Reserves of black coal are located along the eastern seaboard in New South Wales and Queensland. Brown coal (lignite) is mined in the La Trobe Valley of Victoria. Australia has 25% of the world's reserves of lignite.

Oil and natural gas

Conventional oil extraction involves drilling a well (borehole) into a reservoir rock (Figure 14). Oil and gas are held within the pore spaces of reservoir rocks, e.g. sandstone. As the pore spaces are connected, oil and gas will flow freely up the well under pressure, or can be pumped out.

However, some gas and oil deposits are found in rocks (especially shale) in which the pores are not connected, often at depths of over 1000 m. These are referred to as tight gas (or shale gas) and tight oil. To release them, the rock must be fractured. The fracturing creates small cracks and fissures allowing the gas and oil to escape. The fracturing process using water, sand and chemicals (known as hydraulic fracturing or fracking) involves high pressure forcing the rock to shatter (Figure 15).

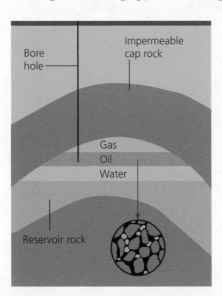

Figure 14 Conventional oil drilling

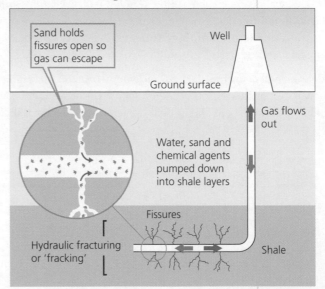

Figure 15 Drilling by fracking

Geothermal energy

This is derived from the heated rocks beneath the surface of the earth. Cold water is pumped down boreholes, heated by contact with the hot underlying rocks, and turned to steam. It is then returned to the surface, where it is used to generate electricity.

Knowledge check 57

Identify one country where geothermal power is used extensively and describe how the power is used.

Geothermal power is generated in a variety of locations around the world such as Iceland and New Zealand.

Hydro-electric power (HEP)

HEP is a clean and efficient renewable resource generated by using the power of running water to turn the turbines. This can occur at natural waterfalls, but more usually involves the construction of a dam to impound a lake. This creates a head of water, determined by the height difference between the water intake pipe and the turbine room, which produces the pressure required. Conditions favouring HEP schemes are stable underlying geology, an impermeable drainage basin, high and reliable rainfall and mountainous areas with narrow gorge-like valleys.

Many of the most suitable sites are often remote from markets, as recent improvements in technology allow transmission over distances in excess of 800 km. HEP only contributes a small proportion of total energy, but it is an important source in parts of the world with few oil, gas and coal reserves. It is a major power resource for some developed countries such as Sweden and Switzerland.

Solar energy

Solar energy is the most abundant of all sources of power. Increasing numbers of property owners in the UK have been installing solar panels on their roofs to provide domestic electricity, and the pressures to develop large commercial solar farms are growing. A solar farm is a large array of interconnected solar panels that capture sunlight and use its energy to generate electricity. Any area of the world that has high levels of insolation (irradiation) is a suitable area.

Australia has the highest solar radiation per square metre of any country in the world, receiving an average of 58 million PJ (peta-joules) of solar radiation per year, yet its current use of solar energy is relatively low (see Table 24).

Wind power

Wind power is the renewable source of energy that is being developed commercially most on a large scale. For it to be efficient, a wind turbine should be located at a point where wind is both regular and strong. Suitable sites include hilltops and coastal areas. It is common to find many turbines grouped together to create a 'wind farm'.

The UK has a lot of potential for wind power, both on land (onshore) and on the sea (offshore), and it has a helpful time profile: wind is strongest in winter and usually builds up to an afternoon peak which fits well with the demand for electricity across the country.

Energy supplies in a globalising world

Competing national interests

In general, the developed world (Western Europe, Japan and USA) has a high oil and natural gas dependency, particularly from countries in the middle east. Equally, the emerging economies of Brazil, China, India and South Africa depend on fuel imports, as indeed do many developing countries. A recent feature has been the move towards resource nationalism where, for example, the USA has effectively banned the export of oil and gas. On the other hand, there are energy surplus countries such as Canada, Iran, Saudi Arabia and Russia, which benefit from the export of oil.

In the USA, energy security is a top national priority. While it is still importing some oil, it is developing its resources of shale oil and gas, and exploring deep water resources such as those in the Arctic. Fracking has had a major impact on the energy security of the nation – it is now almost self-sufficient in oil. Much of the USA's increased oil production has come from fracking in fields such as the Bakken oilfield in North Dakota. There have, however, been some environmental costs – much of the oil is taken away by truck. A pipeline (the Dakota Access Pipeline) is currently under construction, but this is proving to be highly controversial as it passes through Native American land.

Russia is seeking to exploit its oil resources along its vast northern shore with the Arctic Ocean. Melting ice has opened up new transit routes, and revealed previously inaccessible oil and mineral deposits. Russia's view of the outside world is coloured by a deep sense of insecurity and so it is keen to develop new resources and to protect them. Russia is interested in the Arctic for a number of reasons:

- The Arctic contains an estimated 30% of the world's undiscovered natural gas and 15% of its undiscovered oil reserves.
- The northern sea route from eastern Asia to Europe via the Arctic Ocean provides an economic opportunity for developing infrastructure in northern Russia.

Consequently, militarising the Arctic will be a key imperative for the Russian government throughout the coming decade. Bases in the Arctic are being reactivated – the airstrip on the archipelago of Novaya Zemlya is being renovated, and part of the Russian Northern Fleet is based on the island chain. The Arctic Ocean represents a significant area of competing national interests regarding future energy supplies.

The role of TNCs

TNCs have a significant role in energy supplies in a globalised world. Five of the world's ten largest non-financial TNCs are oil and gas companies: Royal Dutch Shell, Exxon Mobil, Total, BP and Eni S.p.A. These companies are heavily involved in all aspects of oil production, from initial exploration to extraction to refining and often retailing.

BP

The TNC BP operates in over 100 countries, ranging from the Gulf of Mexico, Egypt, Iraq, Indonesia, Jordan, Russia, Angola and Azerbaijan. It has huge reserves of capital and often works in partnership with a range of other TNCs, such as the Russian company TNK in Russia, Southern Oil in Iraq, and the China National Petroleum Company (CNPC). The company has helped to fund a pipeline linked to deepwater drilling in the Gulf of Mexico. Oil is also refined in BP's own refineries, seven of which are in Europe. In the UK, the company has thousands of garage outlets, making it the second biggest retailer (non-supermarket) of fuel.

In recent years, BP came to the world's attention on 20 April 2010, when an offshore drilling rig (the Deepwater Horizon) in the Gulf of Mexico exploded, and sank two days later. The event killed 11 people and caused a massive oil spill which threatened the coast of Louisiana, Mississippi, Alabama and Florida. Since then, and following strong criticism from President Obama, the company has been keen to 'clean up its act', both literally and metaphorically.

Exam tip

You could investigate the controversial plans to 'open up' the Arctic National Wildlife Reserve (ANWR) to oil and gas exploration – Alaska, USA.

Making connections

This section is connected to Global governance – the Global commons.

Exam tip

This is again a highly topical area of study. Try to stay up to date with this important topic.

Environmental impacts of a major energy resource development: Northern Alberta, Canada

Exam tip

Study the area of Northern Alberta in an atlas or an online source, e.g. Google Earth.

Introduction

The Canadian tar sands of northern Alberta have huge reserves of crude oil. Tar sands consist of bitumen (heavy viscous oil which is solid at normal temperatures), mixed with clay, sand and water. Deposits close to the surface are strip mined by huge hydraulic and electrically powered shovels. Deeper deposits (around 80% of total reserves) are extracted in situ. Steam is injected to melt the bitumen which then pools near the surface. Most of the major oil TNCs, including Shell, BP, Chevron and Exxon, are involved in the development of tar sands, including those at Athabasca, Peace River and Cold Lake.

Exploitation of the tar sands is highly controversial. Federal and provincial governments believe that the economic benefits outweigh any environmental and social costs. TNCs also support the development, viewing the tar sands as an important long-term investment. However, many scientists, conservationists and First Nation (i.e. indigenous) people strongly oppose development on environmental, social and cultural grounds.

Economic benefits

As the world's more accessible oil reserves deplete, higher cost resources like the tar sands become commercially attractive. As a result, Canada is now a net exporter of oil and self-sufficiency gives Canada greater energy security. At a local scale, mining and processing generate around 100,000 jobs in northern Alberta. Shell directly employs nearly 3,000 workers in Alberta, half of them from First Nation groups. In addition, many First Nation people benefit from service contracts and land agreements with the oil companies.

Environmental costs

The negative environmental effects of the tar sands development are unequivocal. The Athabasca River downstream from the mining areas is of crucial importance to the economy and culture of native Chipewyan and Cree Indians. Water abstracted from the Athabasca River (for separating bitumen from waste materials) has reduced river flow, increasing the pressure on wildlife. The river is unnavigable in summer and restricts the access of native people to traditional fishing, hunting and trapping grounds.

Extracting oil from the tar sands is a very dirty business. The area was originally a pristine forest wilderness of northern Alberta. Strip mining has resulted in huge damage to the environment. Primary boreal forest and lakes are replaced by a wasteland, devoid of vegetation, mining pits up to 100 m deep, and tailings ponds filled with toxic waste.

Scientists believe that mining has released heavy metals such as cadmium and mercury into the Athabasca River. These pollutants present a serious risk to food chains, water supplies and human health. Local people say that the Athabasca River has become more turbid due to increases in suspended sediment, which has contaminated drinking water and damaged fish stocks and other aquatic life. Abnormally high levels of cancer have been reported among First Nation people in northern Alberta close to the tar sands operations.

Shell has responded to criticism of excessive water abstraction from the Athabasca River by pointing out that it recycles much of the water that flows to the tailings ponds. Oil companies operating in Canada are required by law to restore land disturbed by mining to its original state. However, reclamation is failing to keep pace with new development and less than one-eighth of the area disturbed by mining has been reclaimed.

Strategies to increase energy supply

Oil and gas exploration

The geological basis of hydraulic fracturing (fracking) has been outlined earlier. This section will look at arguments both for and against its development in the UK.

Arguments for

The business community and the UK government have been keen to emphasise the potential economic benefits of developing shale gas. In 2013, the Institute of Directors suggested that 'shale gas could represent a multi-billion pound investment, create tens of thousands of jobs, reduce imports, generate significant tax revenues and support British manufacturing'. It estimated that shale-gas production in the UK could attract investment of £4 billion per annum and support up to 74,000 jobs.

Arguments against

The pressure group Friends of the Earth describes fracking as 'unconventional, unnecessary and unwanted' not least because 'fracking for shale gas keeps us hooked on fossil fuels instead of moving towards an energy system based on energy saving and renewable sources'.

National opposition groups including 'Frack Off' and local groups including 'Residents Action on Fylde Fracking' have emerged where fracking is planned. The latter describes itself as 'a group of local residents who are calling for an immediate halt to gas extraction on the Fylde coast pending thorough evaluation of the risks'.

All such groups list a wide range of concerns including:
- lack of regulation
- health risks from air and water pollution
- increased industrial traffic flows
- the threat of small earthquakes and land subsidence
- damage to an area's tourism businesses

UK government response

The government's initial response was that shale gas development 'must be done in partnership with local people' and that it wants 'to encourage a shale industry that is safe and doesn't damage the environment'. However, following a series of small seismic events in the area around the Preston New Road (Fylde) exploratory site in Lancashire, the government announced in November 2019 the halting of all fracking activities in the UK.

Nuclear power

This involves the use of radioactive energy produced by nuclear reaction. The heat released by the reaction is used to produce steam which drives turbines to produce

Exam tip

Some believe that this conflict is not at an end – keep an eye on recent developments.

electricity. Uranium is processed, enriched and converted to uranium dioxide, which is used in the reactor. This undergoes nuclear fission, which releases large amounts of heat. Only small amounts of uranium are needed to produce a given output of heat compared with other forms of fuel. The plants are expensive to construct and decommission, and there is a major concern relating to the disposal of the waste material.

Supporters say that nuclear power must be seen in the context of world energy needs; it is unlikely that demand can be met from renewable sources and yet other sources such as coal, oil and gas are finite. There is no shortage of uranium and operating costs are very competitive. However, there are questions of safety and environmental damage, but so far nuclear plants have a good record in relation to other forms of electricity production and they produce less CO_2 and SO_2. Disposal of waste is a problem because it is radioactive and requires specialised treatment and there is strong public concern that some contamination will occur in transit or at the point of disposal (see page 127).

Renewable sources

Renewable sources of energy are growing in importance. The UK government is committed to net zero carbon emissions by 2050 and hence renewables will have a crucial role to play in the UK energy mix in the coming decades. This mix currently includes solar power and wind power.

Solar power

Solar energy is the most abundant of renewable sources of power. A solar farm is a large array of interconnected solar panels that capture sunlight and use its energy to generate electricity. They have no moving parts so they produce very little noise, and they emit no harmful gases.

A number of factors have driven solar farm development around the world:

■ Government support for solar power has often taken the form of feed-in tariffs – fixed prices which must be paid by the electricity companies for the power that solar farm operators feed into the national grid, and tax credits.
■ As the costs of developing and operating solar power stations have fallen and the cost of electricity generation by traditional methods has increased, solar power has achieved grid parity, where electricity generated from solar becomes equal in cost or cheaper than power from the grid.

China, Germany and the USA currently have the largest solar farm capacity.

Wind power

The physical requirements of wind power have been discussed earlier. Of all renewable energy sources, wind power occupies a unique place due to a combination of two attributes:

■ Technological preparedness – wind is the best placed of all existing renewable sources to contribute to the electricity needs of several European countries while simultaneously reducing their CO_2 emissions.
■ It is inherently site specific – making wind turbines strikingly visible additions to often previously undeveloped landscapes.

Knowledge check 58

Describe the current state of nuclear power generation in the UK.

Knowledge check 59

Provide details of one large solar power facility in the world.

In November 2020, the UK government announced plans to harness enough offshore wind power to provide electricity to every home in the UK by 2030 – a total of 40 gigawatts.

Wind power is generated on land (onshore) and at sea (offshore).

Onshore wind

Wind turbines are seen as a boon to many farmers because of the income that they generate through rent. A very large turbine could generate up to £40,000 a year in rent and even small ones provide substantial help to a farmer who is struggling to make a livelihood. Most of the cable connections are buried, so there are rarely additional pylons to see. All onshore wind turbines are accessible and easy to maintain, though regular checks are usually undertaken remotely online. There is some noise as the blades swish past the column and some background hum from the gearbox in the nacelle at the top. However, many people perceive them as a visual intrusion and do regard them as noisy.

Offshore wind

The biggest advantage of offshore wind farms is that they can be extremely large. Several of the largest arrays in the world are around the UK coast, for example in the Thames estuary (e.g. the London Array) and on the west coast near Wales and Cumbria. Each of these arrays has over 100 large turbines. The costs run into many millions of pounds – they are major construction projects, requiring a skilled team of engineers. The problems of anchoring a large turbine to the seabed are considerable and have limited the depth of water in which they can be sited. As windy offshore areas provide a hostile environment, these turbines have to be extremely reliable, with a design life of about 25 years. It is too expensive to get maintenance personnel to visit offshore turbines frequently. This engineering challenge contributes to the greater cost of offshore activity.

> **Exam tip**
>
> Be prepared to evaluate the relative usefulness and impact of sources of wind power.

Strategies to manage energy consumption

The reduction of energy consumption can be achieved by greater efficiency and by using more renewable resources. Reducing demand is also a key part of managing energy consumption. Energy conservation initiatives include:

- smart electricity meters installed in homes
- high levels of insulation in new homes and the promotion of double and triple glazing and loft insulation schemes for older housing
- improving public transport systems, making them easier and cheaper to use. For example, in London, Oyster cards allow access to all forms of public transport
- more sustainable homes, which includes: installing solar panels and ground source heat pumps and ensuring efficient use of energy through high efficiency central heating boilers (in the UK, the sale of gas boilers will be prohibited by 2033); lighting systems with low-energy light bulbs; the landscaping and design of houses to maximise heat from the sun on the south-facing side
- sustainable transport developments including the development of light rail systems, e.g. the Manchester Metrolink and Sheffield Supertram Network

- the development of low-carbon vehicles (electric and hybrid): in November 2020, the UK government announced that from 2030 no cars and vans powered by petrol or diesel will be sold, though hybrid cars will be allowed
- businesses can be made more sustainable through promoting car-sharing schemes, employees working from home (a significant consequence of the coronovirus crisis from 2020), and new buildings with energy-efficient lighting and heating

Other sustainability issues

Acid rain (deposition)

Acid rain (deposition) consists of the wet deposition of sulphuric acid, nitric acid and compounds of ammonium from precipitation, mist and clouds, and the dry deposition of sulphur dioxide, nitrogen oxides and nitric acid. Acid rain leads to direct damage to trees, particularly coniferous trees. It produces a yellowing of the needles, and strange branching patterns. It also leads to the leaching of toxic metals (aluminium) from soils, and to their accumulation in rivers and lakes. This in turn leads to the death of fish. Acid rain is also blamed for damage to buildings, particularly those built of limestone, and to health problems in people, such as bronchitis and other respiratory complaints.

The major causes of acid rain are the burning of fossil fuels in power stations, the smelting of metals in older industrial plants, and exhaust fumes from motor vehicles.

Various solutions to acid rain have been introduced:
- the use of catalytic converters on cars to reduce the amount of nitrogen oxides
- burning fossil fuels with a lower sulphur content
- replacing coal fired power stations with nuclear power stations
- the use of flue gas desulphurisation schemes

Nuclear waste

Nuclear waste consists of both high-level and low-level radioactive material. Nuclear power stations produce high-level waste which consists of used fuel rods and cells which have been removed from the reactor. These are transferred to a reprocessing plant, such as the Thorp plant at Sellafield (Cumbria), where reusable uranium and plutonium are separated out to leave radioactive waste. This is currently put into steel-clad glass containers, which are then stored at Windscale in Cumbria.

Low-level waste includes clothing and materials used in hospitals where exposure to radium and X-rays can present a low-level risk. These materials are disposed of in controlled chemical and radiation dumps and may be buried in suitable geological structures.

In the UK, the Nuclear Decommissioning Authority (NDA) is charged with investigating sites for the long-term burial of all radioactive waste. As yet no site has been chosen. Any suitable site would need to be deep (over 200 m), geologically stable, near to appropriate transport facilities and acceptable to local populations. Some may argue that the combination of these requirements make the quest impossible.

Knowledge check 60

Provide details of how one city is seeking to manage energy consumption and its impact.

Exam tip

The enhanced greenhouse effect is also stated in the specification. You should revisit your work on the carbon cycle and climate change.

Mineral security: iron ore

Iron is the basic ingredient for steel. Steel is a very useful metal because it can be mixed with other metals to make a variety of 'alloys' which are long-lasting and able to be easily shaped into products such as cars, household appliances, building frames, bridges, railway lines, food cans, tools, nails and pins. The world relies on iron (as steel) to make many of the items we need for living in the twenty-first century.

Sources and distribution

Iron ore (chemical symbol Fe) tends not to occur in isolation – it is combined with other elements in rocks. For example, **haematite** (Fe_2O_3 (70% Fe)) is a red ore of iron and is responsible for the red colour in many rocks, and the deep red sands of the Australian deserts. Another form of iron ore is **magnetite** (Fe_3O_4 (72% Fe)).

Most of the world's important iron ore resources occur in iron-rich sedimentary rocks known as banded iron formations (BIFs), which are almost exclusively of Precambrian age (greater than 540 million years old). BIFs occur on all continents. Haematite ore dominates world production and is sourced mainly in China, Australia and Brazil. Magnetite ore is increasing in production, and has contributed to the increase in Australia's resources of iron ore, particularly in Western Australia and South Australia.

China dominates iron ore production with 45% of world production followed by Australia with 22% and Brazil with 10%. Iron ore reserves in Australia are the highest in the world with over 25 billion metric tonnes of iron content (2016). Over 90% of Australia's identified resources/reserves exist in Western Australia, particularly in Hamersley Province. Australia contributes to over half of the world's iron ore exports.

Brazil is also a major producer and exporter. Some of the world's largest mines are found in Brazil, such as the Carajas iron mine in the north of the country. Carajas has approximately 7.2 billion metric tons of proven iron ore reserves. However, its proximity to the Carajas National Forest has prompted environmental concerns.

End uses, components of demand and role in industry

Although iron in its cast form has some specific uses (e.g. pipes, fittings, engine blocks), its main use is to make steel. Steel is strong, durable and extremely versatile. Many different kinds of steel consist of iron with the addition of small amounts of carbon (usually less than 1%) and other metals to form different alloys (e.g. chromium for stainless steel). Most of the additional elements in steel, e.g. chromium, manganese, nickel, molybdenum or titanium are added deliberately in the steelmaking process within a blast furnace. By changing the proportions of these additional elements, it is possible to make alloy steels suitable for a great variety of uses. These alloys give steel special properties such as electrical resistance, and are resistant to wear, rust, impact, shock or expansion when heated. Steel can also be shaped and coated with tin, zinc or paint to help protect it from rusting.

Knowledge check 61

Describe how your chosen mineral is used.

Physical geography and ore occurrence

These are best studied in the context of a case study: **Hamersley Province (Western Australia)**.

Here there are three main types of deposit:

- iron oxide enrichments within BIFs
- iron oxides deposited along ancient, mainly Tertiary age, river channels (palaeochannels)
- iron oxide deposits formed from the erosion of existing ore bodies (detrital iron ore deposits)

The BIF enrichment deposits are the most important with regard to production. The iron content of these ores varies widely and most deposits need to have an average grade of more than 55% Fe for mining to be commercially viable. The palaeochannel deposits, composed of limonite, are the next in importance and are prized for the low level of impurities such as phosphorus. They are not as rich in iron as the BIF enrichment ores. Detrital iron ore deposits, including scree deposits, are found downslope of the BIF enrichment deposits from which they have been eroded. They are usually easily recovered and have a grade of between 40%–50% Fe.

All the major Western Australian iron ore mines are open cast. The ores from the major mines are hauled from working faces to crushing and screening plants using trucks that can carry over 300 tonnes. The ore is then transported for further concentration to port sites in trains consisting of up to three locomotives and over 250 wagons. Trains of this size are over 2 kilometres long and contain loads in excess of 25,000 tonnes.

Beneficiation includes all the processes that increase (upgrade) the iron content of an ore by removing impurities, and includes pelletising. Many of the iron ore mines employ some simple form of beneficiation to improve the grade and properties of their ores. At some mines, including Mount Tom Price, Mount Whaleback and Christmas Creek, ore processing facilities have been constructed to enable beneficiation on site of low-grade iron ores.

Environmental impacts

These are best studied in the context of a case study: **Bento Rodrigues, Brazil**.

In November 2015, two dams collapsed at an iron ore mine in southeastern Brazil. The dam was owned by Samarco. More than 62 million m^3 of wastewater, much of it sludge, was unleashed with catastrophic consequences. The sludge wiped out a number of villages, including Bento Rodrigues, causing the deaths of 12 people and many more were declared missing.

The wastewater further restricted access to clean water for up to half a million people for both drinking and for irrigation. By 23 November, the contaminated waters covered a 600 km stretch of the Rio Doce river and entered into the Atlantic Ocean on the coast of Espírito Santo State, killing significant numbers of fauna. People were concerned that they might also threaten the Comboios Nature Reserve, a protected area for the endangered leatherback turtle.

Experts, including the UN Human Rights Council (UNHRC) who visited Brazil after the disaster, attributed the bursting of the dam to a severe failure in the preventative approach by the managing companies, as well as inadequate enforcement of regulation in Brazil's mining. Samarco was fined US$265 million, though the Brazilian government sought compensation of more than $5 billion.

In Barra Longa, a small town 70 km from the dam, where some houses were swamped in 2 m of mud, residents said the responses of Samarco and the local government were slow. The mud was also found to contain high levels of arsenic and other toxic materials. It is estimated that the region could take anything from 10 to 50 years to recover, if it actually does. Thousands of fishermen dependent on the river for their livelihood, and several fishing unions, have launched lawsuits against Samarco.

In the past 20 years, there have been at least four dam bursts in the Minas Gerais region, although none as serious as the Bento Rodrigues disaster. A report by the UNHRC stated that, of the similar 750 dams in the Minas Gerais region, 40 were 'at risk' of failure. However, the government of Brazil states that, like most developing countries rich in natural resources, it needs large projects to create investment, wealth and jobs and, as a result, regulations may be overlooked. It also appears that the political will to implement sufficient regulations is weak.

Sustainability issues

Table 25 Sustainability issues for mineral resources and possible solutions

Sustainability issue	Possible solutions
Use of land – the land area used is larger than the mine itself. This can result in considerable loss of habitats	Restoration and rehabilitation plans – involving replanting and reintegration of plant and animal species
Pollution – visual, noise and air	Use of baffle mounds to absorb some sound and also to act as visual screens Dust can be reduced by using water spraying
Water turbidity – fine deposits in water courses can block sunlight for aquatic plants and choke some	The use of holding ponds or lagoons to enable the suspended load to settle
Leaching of toxic materials – some toxic minerals mobilised by mining can kill both flora and fauna	The use of limestone filters to neutralise acid waters from mines This can also immobilise some toxic metals
Disposal of spoil and wastes – spoil heaps can have the potential for mass movement to create flows and slides	Good-quality drainage systems that are constantly monitored to reduce the water content of the spoil heaps
Sustainable trade patterns – there are concerns that some countries concentrate more on meeting demand from other countries at the expense of the environment	Proposal for a Green Trade Alliance (GTA) between producers and importers. This would enforce protectionist trade to encourage non-GTA countries to comply with environmental standards
Recycling is a key theme – over 75% of steel is recycled in the world today. However, this option is not always easy to adopt (transport and labour costs are high), and the use of mixed minerals in products makes recycling difficult	Government legislation can enforce recycling (e.g. the EU has a number of directives). Financial penalties also exist, e.g. landfill taxes in the UK. Promotional campaigns can encourage more recycling, re-using and redesigning

Exam tip

Be prepared to evaluate the relative importance of these issues and the solutions to them.

Resource futures

Water futures

The challenges

Growing pressure on freshwater resources from climate change and from population and economic growth poses a growing risk to many areas. It is estimated that global water requirements will grow from 4,500 billion m^3 in 2018 to 6,900 billion m^3 in 2030. This is 40% above the current level of accessible, reliable supply. It is also estimated that one-third of the world's population, concentrated in developing countries, will live in basins where this deficit is larger than 50%.

The drivers of this challenge are fundamentally tied to economic growth and development. Agriculture accounts for approximately 70% of global water withdrawals today, and without efficiency gains this figure will increase to 4,500 billion m^3 by 2030. Industrial withdrawals account for 16% of today's global demand, growing to a projected 22% in 2030. The growth will come primarily from China (where industrial water demand in 2030 is projected at 465 billion m^3, driven mainly by power generation), which accounts for 40% of the additional industrial demand worldwide. Overall global demand for water for domestic use will decrease as a percentage of total consumption, from 14% today to 12% in 2030, although it will grow in the economically emerging countries.

While the gap between supply and demand will have to be closed, the question is *how*. Historically the focus for most countries in addressing this water challenge has been to develop additional supply, in many cases through energy-intensive measures such as desalination. However, desalination is vastly more expensive than traditional surface water supply infrastructure (dam-building), which in turn is often much more expensive than efficiency measures, such as better irrigation practices in agriculture.

Solutions

Solutions in any particular area could utilise a combination of three fundamental ways to close the demand–supply gap:

- increasing supply
- improving water productivity and efficiency
- reducing withdrawals

The first two are technical solutions, and will require capital investment of approximately $60 billion by 2030 to close the expected water resource availability gap. Agricultural productivity is a fundamental part of the solution. Increasing 'crop per drop' can be achieved through a combination of improved efficiency of water application and net water gains through higher crop yields. These include the familiar technologies of improved water application, such as increased drip and sprinkler irrigation. Efficiency in industry and municipal systems is similarly critical.

The availability of finance is crucial. There is wide agreement that water management has suffered from chronic underinvestment. Financial institutions are likely to be an important factor in making up this shortfall. However, the measures that require the most capital in each country are different, e.g. municipal leakage reduction in China, water transfer schemes in South Africa and investment in drip irrigation in India.

Innovation in water technology – from supply (such as desalination) to industrial efficiency to agricultural technologies (such as irrigation controls) – could play a major role in closing the supply–demand gap.

It is likely that water will be an important investment and development theme for national governments, multinational agencies and private financial institutions in the coming decades. Although affordable solutions are available to close the projected water supply–demand gaps for most countries and regions, political barriers, lack of awareness and poor governance may stand in the way of implementation.

Exam tip

Be prepared to evaluate the relative importance of these challenges and the solutions to them.

Energy futures

Primary energy production and consumption

When looking at energy production and consumption over the next two decades, three questions could be asked:

- Will the world have sufficient energy to fuel continued economic growth?
- Will that energy be secure?
- Will that energy be sustainable?

For the first question, the answer is probably 'yes'. It is projected that global energy consumption will rise by 40% by 2035, with 95% of that growth coming from the emerging economies. That growth rate is slower than that seen in previous decades, largely as a result of increasing energy efficiency. Trends in global technology and investment suggest that production will be able to keep pace. New energy forms such as shale gas and renewables will account for a significant share of the growth in supply.

On the second question of security, the key consideration is one of sufficiency. The USA is on a path to achieve energy self-sufficiency, while import dependency in Europe, China and India will increase. Asia will become the dominant energy-importing region, with India surpassing China in terms of its increase in demand. Russia will remain the leading energy exporter, and African countries will become increasingly important suppliers. While it will remain a key energy supplier, the middle east is likely to see relatively static exports.

On the third question of sustainability, it is accepted that global carbon dioxide emissions will rise significantly, with most of the growth coming from the emerging economies. There are, however, some positive developments: emissions growth will slow as natural gas and renewables gain market share from coal and oil.

Electricity generation

By 2035, it is estimated that 45% of primary energy will be converted into electricity. At the global level, coal will remain the largest source of electric power till 2035, although in the OECD countries, coal will be overtaken by gas. Carbon-free sources (renewables, HEP and nuclear) will increase their combined share of electric power generation to 40% (possibly larger) by 2035.

Energy trade

Regional energy imbalances, i.e. production minus consumption for each region, suggest that trading relationships will change significantly by 2035. North America has already switched from being a net importer of energy to a net exporter.

Meanwhile, Asia's need for imported energy will continue to expand; by 2035, Asia will account for 70% of net imports – and nearly all of the growth in trade. Among exporting regions, the middle east will remain the largest net energy exporter, but its share will fall to 40% in 2035 (from 45% in 2018). Russia is likely to remain the world's largest single energy-exporting country.

Mineral ore futures

There will continue to be concerns over the security of future sources and supplies of key minerals. New deposits are constantly being found, and previously known reserves that were originally thought to be too expensive to access are being developed with improvements in technology. Another interesting aspect is that the demand requirements for minerals are changing as new minerals become necessary for new industrial processes, such as the rare earth elements (REEs) discussed earlier. The minerals antimony (used in batteries and microelectronics), and tungsten (used for its hardness, and especially so in military products), are important 'new' minerals.

Technological developments have been crucial in the exploration of new deposits. Such techniques include seismic surveys (expanding their use from earthquake measurement), magnetometry (used to discover new sources of iron ore), gravimetry (used to discover igneous rocks which may contain REEs) and remote sensing (the use of satellite and infrared imagery). Technological advances in deep-sea exploration using unmanned submarine vessels have also assisted here.

Technology has also improved the mining process itself. Some of the excavating equipment used is massive in scale, and the ability to mine huge deposits of a lower grade has made their exploitation much more cost-effective.

Economically, mineral exploitation depends on the traditional aspects of supply, demand and price fluctuations. The prices of 'staple' ores such as iron, copper and nickel do not tend to fluctuate much. However, with the increase in the incidence of recycling, the price of steel has fallen in recent years, as evidenced by the current issues facing the UK steel industry caused by a 'glut' of cheap steel on international markets.

Environmentally, the impacts of excavation will continue to manifest themselves on landscapes, but as many of these sites are remote and largely unpopulated, their environmental impact may cause less of a concern. However, when 'mistakes' are made, e.g. sites important to indigenous people being disturbed, negative headlines appear in the media affecting perception of the industry. NGOs are the main drivers of environmental protection and Environmental Impact Assessments (EIAs) are becoming more important prior to approval of new mining projects.

Politically, it is likely that new mining projects will spread to the emerging markets of Latin America and Africa, especially in Brazil, Chile, the Democratic Republic of Congo and South Africa. These will depend on FDI from both western and Chinese TNCs, which will be seen as welcome investment in terms of providing jobs and infrastructure.

Another potentially politically significant area of mineral exploitation not yet utilised is Antarctica. It is predicted that this continent contains huge deposits of minerals. The Antarctic Treaty (ATS) deliberately omitted any reference to Antarctic mineral

Making connections

This section is connected to Global governance – the Global commons (Antarctica).

potential, for fear it would jeopardise the agreement. In 1991, the (Madrid) Protocol on Environmental Protection was negotiated, and this entered into force in 1998, prohibiting all forms of mineral exploitation. The ATS parties, alarmed by criticism from environmental groups and critics in the UN General Assembly, decided on a blanket ban. However, the ban can be revisited in 2048. In the last decade, there have been signs that some parties want to revisit the situation regarding minerals. China and Russia in particular have been accused of consolidating their presence on Antarctica (in the form of scientific stations and infrastructure) for the purpose of strengthening their interest in long-term mineral wealth.

Knowledge check 62

How are the resources of Antarctica being exploited currently?

Case studies

Two case studies are specifically required by the specification:

1 A case study of **either** water **or** energy **or** mineral ore resource issues in a global or specified regional setting to illustrate and analyse theme(s) set out above, and their implications for the setting including the relationship between resource security and human welfare and attempts to manage the resource. Examples include the Chinese South–North water transfer project (water), the Canadian Tar Sands (energy) or the Bento Rodrigues mine (iron ore).

2 A case study of a specified place to illustrate and analyse how aspects of its physical environment affect the availability and cost of water **or** energy **or** mineral ore and the way in which water or energy or mineral ore is used. Examples include Gaza (water), the Kashagan oilfield in Kazakhstan (energy), or Mount Goldsworthy in Western Australia (iron ore).

Summary

After studying this topic, you should be able to:

■ understand the principles of resource development, including classification, evaluation, resource frontiers, resource peaks and Environmental Impact Assessment
■ know and understand the global patterns of production, consumption and trade of energy and ore minerals, together with the global patterns of water availability and demand
■ understand the geopolitics of resource issues (for energy, water and mineral ores)
■ appreciate how water security can be maintained by identifying sources, understanding links to physical geography, and recognising the various strategies to manage water supply and consumption
■ appreciate how energy security can be maintained by identifying sources, understanding links to physical geography, and recognising the various strategies to manage energy supply and consumption
■ appreciate how mineral security can be maintained by identifying sources, and understanding links to physical geography
■ understand that each of the above generates sustainability issues, including environmental impacts, that may create areas of conflict
■ analyse futures for each of water, energy and mineral ores

Questions & Answers

About this section

In this section of the book, two sets of A-level questions on each of the content areas are given. For each of these, the style of questions used in the examination papers has been replicated, with a mixture of short answer questions, data response questions, data stimulus questions and extended prose questions. Other than the short, knowledge-based questions, all will be assessed using a 'levels of response' mark scheme to a maximum of four levels.

The sections are structured as follows:
- sample questions in the style of the examination
- mark schemes in the style of the examination
- example student answers at an upper level of performance
- examiner's commentary on each of the above

For A-level geography, all assessments will test one or more of the following Assessment Objectives (AOs):
- **AO1**: demonstrate knowledge and understanding of places, environments, concepts, processes, interactions and change, at a variety of scales.
- **AO2**: apply knowledge and understanding in different contexts to interpret, analyse, and evaluate geographical information and issues.
- **AO3**: use a variety of relevant quantitative, qualitative and fieldwork skills to: investigate geographical questions and issues; interpret, analyse and evaluate data and evidence; construct arguments and draw conclusions.

All questions that carry a large number of marks require students to consider connections between the subject matter and other aspects of geography, or to develop deeper understanding, in order to access the highest marks. The former used to be referred to as **synopticity**, but the new term is now **connections** – so try to think of **links** between the subject matter you are writing about and other areas of the specification. Some questions will target specific links.

For the Human Geography examination paper 2, two **core topics** on each paper (Global systems and global governance, and Changing places) are covered, each worth 36 marks, and the breakdown of the questions per topic is:
- one 4-mark question (AO1) – point marked
- one 6-mark question with data – marked to two levels (AO3) – **response**
- one 6-mark question with data – marked to two levels (AO1/AO2) – **stimulus**
- one 20-mark question requiring an extended prose response marked to four levels (AO1/AO2)

There are also the **options**: the sections testing Contemporary urban environments, Population and the environment and Resource security are worth 48 marks each, and the breakdown of questions per topic is:

■ one 4-mark question (AO1) – point marked
■ one 6-mark question with data – marked to two levels (AO3) – **response**
■ one 9-mark question (sometimes with data: **stimulus**) requiring an extended prose response marked to three levels (AO1/AO2)
■ one 9-mark question requiring an extended prose response marked to three levels (AO1/AO2)
■ one 20-mark question requiring an extended prose response marked to four levels (AO1/AO2)

Note that the latter two questions may have an explicit connection to another part of the specification.

You should allocate 1.25 minutes per mark to answer the questions.

For each question in this book, one answer has been provided towards the upper end of the mark range. Study the descriptions of the 'levels' given in the mark schemes carefully and understand the requirements (or 'triggers') necessary to move an answer from one level to the one above it. You should also read the commentary with the mark schemes to understand why credit has or has not been awarded. In all cases, actual marks are indicated.

Question types

Short answer questions

These questions assess AO1 only, and carry four marks each. You are expected to provide four clear statements (though not necessarily four sentences) which address the question and demonstrate your knowledge of the topic. Your statements may provide examples, or elaboration, but these are not required.

You have about 5 minutes to answer these questions, so your points should be punchy and to the point.

Data-based questions

The examination papers have two types of questions based on data.

Data response questions

These all carry 6 marks and assess AO3 only. In general, simple or obvious statements will access Level 1; more sophisticated statements will access Level 2. Note that knowledge is not required here, so do not try to explain the data – although you may be correct in such statements, you will *not* gain any credit for them.

Here are some general tips about addressing such questions:

■ consider patterns/ranges/trends
■ identify anomalies/countertrends

- manipulate the data (e.g. calculate percentages, or fractions; use qualitative descriptive words) – don't simply 'lift' or copy them
- make connections/draw relationships between the different sets of data provided
- be prepared to question and/or criticise such relationships, or indeed the data provided

You have about 8–9 minutes to answer these questions.

Data stimulus questions

These questions assess AO1 and AO2 in the proportion of 2/4 for 6-mark questions (core topics) and 4/5 for 9-mark questions (optional topics). You should demonstrate that you know the factors that underpin the context of the data provided, but also apply this knowledge to the question given. It is important that you refer to the data provided, but then use it as a stimulus to answer the question. In other words, use the data provided as a 'springboard'. Note the questions often use the phrase '… and your own knowledge'. There is also often an element of assessment or evaluation in the question (such as the use of the command word 'assess') – this is the AO2 part of the question, and it must be addressed.

You have about 8–9 minutes (6 marks), or 10–12 minutes (9 marks) to answer these questions.

The 9-mark extended prose questions

These questions assess AO1 and AO2 in the proportion of 4/5 and are *only* found within the **Optional** elements. You should demonstrate that you know the factors that underpin the context of the question, but also then apply this knowledge to the question given. There is usually an element of assessment or evaluation in the question – this is the AO2 part of the question, and it must be addressed.

At least one of these 9-mark questions across the examination as a whole will connect the study of geography across this specification (sometimes referred to as the synoptic question). There is no pattern as to where such questions fall – they may be on either Paper 1 or Paper 2, or both.

You have about 12–13 minutes to answer these questions. You should aim to write about 250–350 words.

The 20-mark essay questions

You are required to write *three* evaluative essays on each examination paper. Each essay should be completed in around 25 to 30 minutes. In general, this represents around 2–3 pages of average-sized handwriting, i.e. somewhere between 400 and 600 words. These evaluative essays should incorporate an introduction and a formal conclusion, with several paragraphs (3–5) of argument in between.

These essays are assessed using a generic mark scheme such as the one below. Study this carefully to see what is needed to move from one level to the next.

Level/mark range	Criteria/descriptor
Level 4 (16–20 marks)	■ Detailed evaluative conclusion that is rational and firmly based on knowledge and understanding, which is applied to the context of the question (AO2). ■ Detailed, coherent and relevant analysis and evaluation in the application of knowledge and understanding throughout (AO2). ■ Full evidence of links between knowledge and understanding and the application of knowledge and understanding in different contexts (AO2). ■ Detailed, highly relevant and appropriate knowledge and understanding of place(s) and environments used throughout (AO1). ■ Full and accurate knowledge and understanding of key concepts and processes throughout (AO1). ■ Detailed awareness of scale and temporal change, which is well integrated where appropriate (AO1).
Level 3 (11–15 marks)	■ Clear, evaluative conclusion that is based on knowledge and understanding, which is applied to the context of the question (AO2). ■ Generally clear, coherent and relevant analysis and evaluation in the application of knowledge and understanding (AO2). ■ Generally clear evidence of links between knowledge and understanding and the application of knowledge and understanding in different contexts (AO2). ■ Generally clear and relevant knowledge and understanding of place(s) and environments (AO1). ■ Generally clear and accurate knowledge and understanding of key concepts and processes (AO1). ■ Generally clear awareness of scale and temporal change, which is integrated where appropriate (AO1).
Level 2 (6–10 marks)	■ Some sense of evaluative conclusion partially based upon knowledge and understanding, which is applied to the context of the question (AO2). ■ Some partially relevant analysis and evaluation in the application of knowledge and understanding (AO2). ■ Some evidence of links between knowledge and understanding and the application of knowledge and understanding in different contexts (AO2). ■ Some relevant knowledge and understanding of place(s) and environments, which is partially relevant (AO1). ■ Some knowledge and understanding of key concepts, processes and interactions, and change (AO1). ■ Some awareness of scale and temporal change, which is sometimes integrated where appropriate; there may be a few inaccuracies (AO1).
Level 1 (1–5 marks)	■ Very limited and/or unsupported evaluative conclusion that is loosely based upon knowledge and understanding, which is applied to the context of the question (AO2). ■ Very limited analysis and evaluation in the application of knowledge and understanding; lacks clarity and coherence (AO2). ■ Very limited and rarely logical evidence of links between knowledge and understanding and the application of knowledge and understanding in different contexts (AO2). ■ Very limited relevant knowledge and understanding of place(s) and environments (AO1). ■ Isolated knowledge and understanding of key concepts and processes (AO1). ■ Very limited awareness of scale and temporal change, which is rarely integrated where appropriate; there may be a number of inaccuracies (AO1).
Level 0 (0 marks)	■ Nothing worthy of credit.

Command words

Command words are the words and phrases used in exams and other assessment tasks that tell students how they should answer the question. The following high-level command words could be used:

Analyse (when used for extended prose questions) Break down concepts, information and/or issues to convey an understanding of them by finding connections and causes and/or effects.

Analyse (when used for 6-mark AO3 questions) In this context, this command requires students to interface with the data and deconstruct the information (see above).

Assess Consider several options or arguments and weigh them up so as to come to a conclusion about their effectiveness or validity.

Compare Describe the similarities and differences of at least two phenomena.

Evaluate Consider several options, ideas or arguments and form a view based on evidence about their importance/validity/merit/utility.

Examine Consider carefully and provide a detailed account of the indicated topic.

Explain/Why/Suggest reasons for Set out the causes of a phenomenon and/or the factors which influence its form/nature. This usually requires an understanding of processes.

Interpret Ascribe meaning to geographical information and issues.

To what extent Form and express a view as to the merit or validity of a view or statement after examining the evidence available and/or different sides of an argument.

■ Questions: Set 1

Global systems and global governance

Question 1

Explain one reason why the global shift of industry has had negative impacts on some people in the developed world.

(4 marks)

> 1 mark per valid point.

Student answer

One reason why the global shift of industry has had negative impacts on some people in the developed world is an increase in unemployment ✓. The manufacturing industry has shifted from the West (e.g. Rust Belt, USA) to areas such as China where labour/production costs are lower ✓. For example, in Detroit, USA, over 60,000 jobs in the motor vehicle industry were lost ✓ and this then led to a spiral of decline – a negative multiplier effect which led to the closing down of retail stores in Detroit ✓, land values plummeting, rising crime levels and 'white flight' as the white population left ✓.

4/4 marks awarded The student provides several correct statements.

Question 2

Figure 1 shows trade between China and Africa (2002–2018), and Figure 2 shows trade between the USA and China (2014–2018). Analyse Figures 1 and 2.

(6 marks)

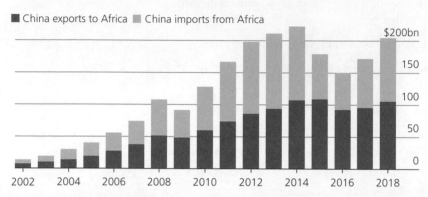

Figure 1 Trade between China and Africa (2002–2018)

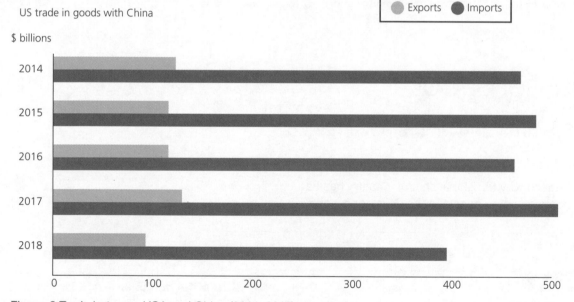

Figure 2 Trade between USA and China (2014–2018)

Level 2 (4–6 marks)

AO3 – Clear analysis of the quantitative evidence provided, which makes appropriate use of evidence in support. Clear connection(s) between different aspects of the evidence.

Level 1 (1–3 marks)

AO3 – Basic analysis of the quantitative evidence provided, which makes limited use of evidence in support. Basic connection(s) between different aspects of the evidence.

Student answer

China's trade with the USA involves a huge amount of money, usually over $600 billion a year, although there was a fall in this during 2018, where it fell to $500 billion. This amount of trade is three times more than that with the whole continent of Africa, where the total amount of trade only once exceeded $200 billion, in 2014.

A big difference shown in the data is that trade with Africa has steadily increased up to 2014, although there were some fluctuations after this date. Similar fluctuations in this time period also occurred between China and the USA, so perhaps the differences are not that clear after all. We don't have the similar data before 2014 for the USA/China trade.

In most years, China has had a trade deficit with Africa, importing more than exporting. However, this has changed in the last four years as exports have been larger than imports – a trade surplus. On the other hand, China has always had a huge trade surplus with the USA (therefore a trade deficit for the USA) – over $300 billion every year.

It is fair to say that trade between China and Africa is more balanced than that between USA and China.

6/6 marks awarded Although the two sets of data may initially appear to be straightforward, the task of analysing these data is more complex. The date periods are different, and the relative 'directions' of the information provided are opposite to each other. Nevertheless, for the most part, the student makes a very good attempt at analysis. In the first paragraph, he/she makes two good quantitative and comparative points.

The second paragraph is a little more confusing, but the student does manage to salvage the point he/she is trying to make.

The third paragraph is excellent, making good qualitative and quantitative points.

The concluding sentence demonstrates full understanding of the outcomes of the data.

Question 3

Figure 3 shows the composition of the Arctic Council, the UN-sponsored organisation responsible for the governance of the Arctic. Using Figure 3, and your own knowledge, assess similar strategies that exist to govern and protect Antarctica.

(6 marks)

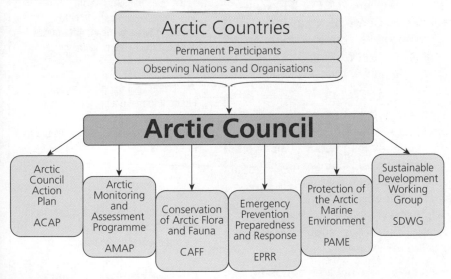

Figure 3 Composition of the Arctic Council

Level 2 (4–6 marks)

AO1 – Demonstrates clear knowledge and understanding of concepts, processes, interactions and change.

AO2 – Applies knowledge and understanding to the novel situation, offering clear evaluation and analysis drawn appropriately from the context provided. Connections and relationships between different aspects of study are evident with clear relevance.

Level 1 (1–3 marks)

AO1 – Demonstrates basic knowledge and understanding of concepts, processes, interactions and change.

AO2 – Applies limited knowledge and understanding to the novel situation, offering only basic evaluation and analysis drawn from the context provided. Connections and relationships between different aspects of study are basic with limited relevance.

Student answer

As with the Arctic Council, the Antarctic Treaty System (ATS) was established by the United Nations in 1959 to designate Antarctica as a zone of peace. Membership of the ATS was open to all countries, and so there was not an arrangement of permanent members and observing members as in the Arctic Council. The Antarctic was identified as a place where science was pivotal for cooperation, and as a result all sovereignty claims were suspended.

The ATS applied to all areas below 60 degrees south and so covered not just the ice sheet mass, but also the sea areas around it. As with the CAFF and PAME in the Arctic, the Convention for the Conservation of Antarctic Marine Living Resources (CCAMLR) was established in the 1980s with the Antarctic Convergence as the spatial limit for living resource management, i.e. flora and fauna. An example of its impact was that the International Whaling Commission (IWC) established the Southern Ocean Whale Sanctuary to protect the whale species in that area.

Both polar regions face resource extraction pressures and so they need careful action planning. From Figure 3, the Arctic has the ACAP. In Antarctica, mining is banned until 2048, although both China and Russia have been at the forefront of pushing for Antarctic resource exploitation. China speaks of 'peaceful exploitation' and aims to expand scientific presence in eastern Antarctica and push for more fish and krill exploitation. This will impact on the sustainability of the continent, something which the Arctic Council has recognised with its Sustainable Development Working Group.

It is clear that there are some similarities in the governance of the two polar regions. It remains to be seen just how successful they will be.

6/6 marks awarded It should be noted that the governance of the Arctic is not a feature of the specification – this question provides the novel situation where students should compare governance in one part of the world (Figure 3) with that taking place in their required area of study, Antarctica. There are clearly some geographical similarities between the two areas. The student makes a number of clear references to Figure 3, and uses it to stimulate his/her answer to the question set. The answer begins with a simple comparison before providing two significant differences between the two organisations.

The second paragraph also uses Figure 3 to make a point regarding the management of fauna – the protection of whales in the seas around Antarctica.

The logic of the argument is continued into the third paragraph, where resource extraction is considered.

Concluding statements are provided, which address the thrust of the question.

Question 4

Evaluate the extent to which TNCs are both a cause and a consequence of globalisation.

(20 marks)

See the generic mark scheme on page 138.

Student answer

Transnational corporations (TNCs), such as McDonald's, Coca-Cola and Nestlé, are large companies that produce or source goods and services internationally and market them worldwide. TNCs are major players in the global economy: the top 500 account for 70% of world trade and generate a large proportion of all foreign direct investment (FDI). Their enormous economic power has been the driving force behind the globalisation in recent decades. It could also be argued that by outsourcing and offshoring, companies inevitably become even more transnational, with mergers and acquisitions of foreign companies automatically leading to the creation of even larger TNCs. So, are TNCs also the outcome of globalisation?

Global inequalities in human welfare are closely linked to economic development. Where economic development has made least progress (e.g. in sub-Saharan Africa) it is hardly surprising that infant mortality rates are high and life expectancy is low. Africa, the poorest continent, has 15% of the world's population yet attracts less than 3% of global FDI. Deterred by poor economic and social infrastructures, political instability and corruption, investment opportunities in much of sub-Saharan Africa remain unattractive to TNCs. Hence, globalisation of large parts of this continent is being manipulated by investment, or rather the lack of it, by TNCs.

TNCs often locate manufacturing plants in developing countries in order to exploit weak labour and pollution laws. The US–Mexico border region has experienced massive economic growth since the 1980s, due largely to investment on the Mexican side of the border by US and European TNCs. This investment mainly comprises branch plants or maquiladoras, which manufacture a wide range of goods including clothes, chemicals and electronics and then export them across the border to the USA.

The main attractions of the border region are low labour costs and proximity to US markets. Another factor here has been the creation of the North American Free Trade Association (NAFTA), an example of free trade between nations. The promotion of free trade, encouraged by national governments and the World Trade Organisation (WTO), facilitates corporate global production lines and TNC activity. Deregulation of capital markets has allowed profits to be moved from place to place and so results in increased TNC activity.

20/20 marks awarded The answer starts with a solid introduction that defines, and exemplifies, TNCs and also engages with one of the themes of the question: cause. This is followed by a brief statement that connects with the second theme – consequences.

The next paragraph, which initially appears to be not related to the question, by introducing some connections to other aspects of the subject, is cleverly brought back to 'cause' in the final sentence.

The next sections again connect with the theme of 'cause' and although not explicitly linked to it, they nevertheless provide a good case study of the spread of globalisation as a result of TNC expansion.

This is then supported by sophisticated points regarding the importance of free trade.

Key aspects of globalisation have been offshoring and outsourcing. In practice, most offshoring involves transferring production from developed countries to developing countries where labour and other costs are lower, and where access to foreign markets is easier. Outsourcing involves subcontracting production to another company, which provides goods and services sold under the TNC brand. Through offshoring and outsourcing, TNCs from developed countries have in effect created more home-grown TNCs in order to undertake their work. For example, companies such as Apple and Primark subcontract the work out to other companies who may in turn operate across regions such as South East Asia – they grow as a consequence of globalisation.

The next paragraph demonstrates thorough knowledge and understanding of the operations of TNCs in the twenty-first century – offshoring and outsourcing – and uses this knowledge to discuss how TNCs are being further developed. This is a very strong paragraph with a coherent sense of assessment.

In contrast, there are advantages created by TNCs, such as the creation of employment, which in turn raises the GNP of the host country. TNCs have also resulted in the growth of global production lines and increased trade in a globalised world, largely through containerisation. Hence TNCs are driving the globalisation of transport systems.

The next paragraph is not as strong but is continuing to explain how TNCs are driving globalisation through transport systems.

In conclusion, the role of TNCs within the global economy is highly significant. The importance of TNCs as drivers of globalisation cannot be overstated. They have encouraged rapid industrialisation in emerging economies such as China, India and Mexico. It is also clear that they themselves have increased in size, numbers and influence in recent years, so indeed it could be argued they are both a cause and a consequence of globalisation.

The conclusion is explicitly aimed at the 'to what extent' part of the task. All elements of the mark scheme for Level 4 have been addressed.

Changing places
Question 1

Explain how resources shape the profile of a place.   (4 marks)

1 mark per valid point.

Student answer

The availability of a mineral resource such as coal or iron ore can lead to a place becoming a mining village ✓. As long as the mining of the resource is economically valuable the place will prosper but once the mine closes a downward spiral can set in ✓. Resources can also be human resources, for example, the location of a university can lead to a place developing as a centre of scientific research ✓. This has occurred on the outskirts of Cambridge, where the Science Park has had a significant effect on the surrounding area ✓.

4/4 marks awarded The student provides four correct statements. Note that the final point is a development (in this case, a named example) of the preceding point.

Question 2

Figure 4 shows two photographs of contrasting places where a religious building is central to the community. Analyse the contrasting characters of the two places shown.   (6 marks)

Figure 4

> **Level 2 (4–6 marks)**
>
> AO3 – Clear analysis of the qualitative evidence provided, which makes appropriate use of evidence in support. Clear connection(s) between different aspects of the evidence.
>
> **Level 1 (1–3 marks)**
>
> AO3 – Basic analysis of the qualitative evidence provided, which makes limited use of evidence in support. Basic connection(s) between different aspects of the evidence.

Student answer

Religious buildings are often at the centre of communities around the world. They are one of the endogenous features that help to determine the character of places. In the case of Photo A, there is a large modern mosque in the midst of some fairly dull concrete buildings in the background and a stilted house community in the foreground. Despite a sense of generally lower overall income in the community, the mosque itself has one very large and a number of smaller gold-covered domes. This may be a sign that much of the wealth in the community has been invested in this building. The same may have been the case when the church in Photo B was constructed, possibly in Victorian times or earlier. The high tower, and its small steeples, will have cost a lot of money in those days.

The areas around the religious buildings show quite large contrasts. In Photo B, there appears to be a village school with some cars parked in the playground. Small schools in a village such as this often have a number of mixed-age classes, but all children in the village will be able to attend. The foreground of Photo A is much more typical of a developing world city where poor people have to find a cheap place to live – in this case on stilted houses with corrugated iron roofs over a river or coastline. The contrast in wealth here is quite stark.

6/6 marks awarded Bearing in mind there are only 7–8 minutes to consider these photographs and answer the question, the opening paragraph is very sophisticated. The student fully understands the phrase 'character of the places', and applies clear evidence from the photographs to make comments on the character of the two places.

The second paragraph is not as strong but the student does attempt to make a clear comparison with some application of knowledge to highlight contrasting characters.

Question 3

Study the information provided in the following extract and Figure 5, which both refer to the town of Whitby in North Yorkshire, England. Using the extract and Figure 5, and your own knowledge, assess how useful qualitative sources of information such as novels and photographs can be in representing a place.

(6 marks)

Information relating to Whitby (North Yorkshire)

Whitby is a small fishing port on the North Yorkshire coast. Nowadays, it is a very popular location for tourists, especially day-trippers. In Victorian times, it was the setting for the novel *Dracula* by Bram Stoker, and this is one of the reasons for its popularity today as fans of the novel visit locations within it, such as the Abbey.

This is how Stoker described Whitby in the novel:

'This is a lovely place. The little river, the Esk, runs through a deep valley, which broadens out as it comes near the harbour. A great viaduct runs across, with high piers, through which the view seems somehow further away than it really is. The valley is beautifully green, and it is so steep that when you are on the high land on either side you look right across it, unless you are near enough to see down. The houses of the old town – the side away from us, are all red-roofed, and seem piled up one over the other anyhow, … Right over the town is the ruin of Whitby Abbey, which was sacked by the Danes, … It is a most noble ruin, of immense size, and full of beautiful and romantic bits.'

Figure 5 Whitby Abbey

Level 2 (4–6 marks)

A01 – Demonstrates clear knowledge and understanding of concepts, processes, interactions and change.

A02 – Applies knowledge and understanding to the novel situation, offering clear evaluation and analysis drawn appropriately from the context provided. Connections and relationships between different aspects of study are evident with clear relevance.

Level 1 (1–3 marks)

A01 – Demonstrates basic knowledge and understanding of concepts, processes, interactions and change.

A02 – Applies limited knowledge and understanding to the novel situation, offering only basic evaluation and analysis drawn from the context provided. Connections and relationships between different aspects of study are basic with limited relevance.

Student answer

Many novels make use of real places as the setting for their stories, sometimes changing the name of the place. They provide us with an alternative perspective of a place, and are of particular use for students of 'place' when they are based, or written, in the past. They provide us with an opportunity to see how the place has changed over time. For example, Thomas Hardy made use of several locations in Dorset for his novels, though he usually changed their names. They provide a useful indication of how these places were in Victorian times.

6/6 marks awarded The opening paragraph is a good one and clearly addresses the question, also making use of an example not provided by the data – this is acceptable for such a data stimulus question.

The extract provides commentary on the features of Whitby in Stoker's time, with the use of adjectives such as 'noble', 'immense', 'beautiful' and 'romantic'. These are clearly from the perspective of the author, and therefore subjective. The extract also provides the description from a person who is visiting the town, an outsider, of features that can only be gleaned from such a visit: '… it is so steep that when you are on the high land on either side you look right across it'. This is true in reality – the old town of Whitby is very much located within a steep-sided valley so that you can only see it from the edge of one of the ridges. The photograph illustrates this too as you can't see the town in the valley.

The second paragraph refers to the data given, and although some valid points are made, there is no explicit link to the question. The use of personal knowledge is acceptable, but once again there needed to be a direct assessment of 'usefulness'.

Photographs are taken in one moment in time, and obviously represent that place at that time. Depending on how they are framed by the photographer they can be used to highlight key features of a place which might stimulate an emotional viewpoint – the photo of Whitby Abbey does illustrate that it is a ruin, but it is difficult to see how beautiful or romantic it is. That has to be in the mind of the individual only.

The third paragraph also strives to address the question, with implicit statements regarding usefulness rather than explicit ones.

So, qualitative materials such as these do help in representing a place, both in reality and in our imagination.

It is clear, though, that the student has interpreted the question correctly – the concluding sentence reinforces this.

Question 4

To what extent have past and present connections shaped the characteristics of one place you have studied?

(20 marks)

> See the generic mark scheme on page 138.

Student answer

Trafalgar Square is known throughout the world as a place of coming together and meeting other people. It lies on the north bank of the river Thames, about 500 m from the river itself and on a marked slope from north to south. This slope marks the first level of river terraces that have been produced over thousands of years by the river in this part of the Thames valley – a strong past connection in the physical geography sense.

Trafalgar Square was the first deliberately planned public square in London. It evolved in stages between 1830 and 1860, and was the capital's first attempt at a grand historical monument commemorating military victory and embodying national pride. The space to the south of the square, on the lower level of the slope, was known as Charing Cross, itself a small open space. This area had previously served as a place of public meetings, proclamations, pillory and execution. The upper part of the slope was originally occupied by the King's Mews, a large area of stables. The demolition of the mews and the nearby buildings began with the project to build a national art gallery. This work began in 1832, sited exactly where the mews had been. A terrace was also constructed on the same level as the gallery, with a series of stone steps descending to the square itself. These steps are still there today. The gallery was completed in 1838, the same year in which Nelson's monument was commissioned. This is a 150 ft granite column with its crowning statue, which was not completed until 1859.

So already by the middle of the nineteenth century, Trafalgar Square had begun to make a name for itself both in the country and beyond. People came to see the splendid constructions, the famous column and the paintings that were beginning to be housed in the gallery. Its characteristics were being shaped by its historical connections.

Further changes were to come. In 1874, a new road to the riverside was created – Northumberland Avenue – named after the large Jacobean house that used to stand on the corner of the square. The final part of Trafalgar Square as we know it came in 1910 with the building of Admiralty Arch as a grand gateway to the Mall. Also by this time two large fountains had been designed by Lutyens and placed

20/20 marks awarded It is difficult to find fault with this answer. The sense of place is strong; the detail provided in the first two paragraphs, both historical and contemporary, is relevant and accurate.

between Nelson's Column and the National Gallery. These added to the attractions of the site, and tourists from all over the world were being attracted to it. Today, New Year revellers like to soak themselves in the fountains.

Today, Trafalgar Square is a landmark enjoyed by Londoners and all visitors alike. It is a lively place, which is often used for a wide range of activities including special events and celebrations: St Patrick's Day, Pride events, Eid and Chinese New Year, and rallies and demonstrations. All of these events bring people together and ensure that its central characteristic, that of being a large meeting space, is maintained. Trafalgar Square is one place that all visitors to London make sure they visit.

Trafalgar Square began its existence as a public meeting place for the people of London. It has now evolved into a meeting place for people across the UK to show their pleasure, as in the case of London being awarded the Olympic Games, or to show their displeasure, as when it was a focal point during the protests for and against Brexit, as well as a place of vigil following terrorist acts around the world. Its characteristics certainly have been shaped by its past and present connections.

> The student has also appreciated the nature of the task – the need to link connections with characteristics – and his/her focus in addressing the task in the third and sixth paragraphs is first-rate. There are also a number of places where the student makes connections (links) to other aspects of geography which are signalled in the specification: physical processes, tourism and conflict – see the first and last paragraphs. All Level 4 criteria have been addressed.

Contemporary urban environments

Question 1

Outline the process of counter-urbanisation. (4 marks)

> 1 mark per valid point.

Student answer

Counter-urbanisation is the movement of people and services from urban areas into smaller towns and villages in more rural areas ✓. Effectively, this means there is a break in settlement between the main city and the area where people are moving to ✓. The push factors from the city are that there is pollution, congestion and it is hard to park, whereas the pull factors to the rural areas are that people believe they are closer to the quiet of nature ✓. This is currently happening in St Ives, Cambridgeshire where the population is now growing rapidly with people who have moved away from cities such as Cambridge, Birmingham and even London ✓.

> **4/4 marks awarded** The student provides four correct statements. Note that the final point is a development (in this case, a named example) of the preceding point.

Question 2

Table 1 provides information about average age, average monthly rent of property and household incomes for selected districts in the city of Vancouver, Canada.

Figure 6 shows a map of the districts of Vancouver.

Analyse the information provided in Table 1 and Figure 6.   (6 marks)

Table 1 Vancouver data (selected districts) (2017)

Area of city	Average age (years)	Average monthly rent ($)	Average total household income ($)
Downtown (DT) West	36	1,400	85,490
West End	51	1,099	54,599
Coal Harbour	36	1,646	85,790
Yale Town	41	1,576	106,989
Point Grey	38	1,280	147,605
Kitsilano	30	1,204	87,825
Kerrisdale	41	1,394	151,582
South Granville	38	1,072	56,512
Fairview	40	1,154	80,073
False Creek	38	1,370	93,073
Mount Pleasant East	49	611	35,904
Main	36	1,051	75,856
Fraser	39	974	69,785
Knight	39	874	57,864
Hastings	41	931	74,368

Source: Vancouver City Council

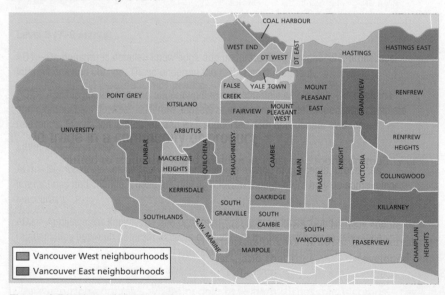

Figure 6 Districts of Vancouver

Level 2 (4–6 marks)

AO3 – Clear analysis of the quantitative evidence provided, which makes appropriate use of evidence in support. Clear connection(s) between different aspects of the evidence.

Level 1 (1–3 marks)

A03 – Basic analysis of the quantitative evidence provided, which makes limited use of evidence in support. Basic connection(s) between different aspects of the evidence.

Student answer

Vancouver has quite a young population, with the average age being 39.5 years. The highest age is 51, West End, where there is one of the lowest incomes in the city. These are perhaps old age pensioners living in an older part of the city. The youngest area (30 years) is Kitsilano, in Vancouver West. Here the income levels are 50% higher than in West End, so possibly we have young adults some with children having better incomes due to higher levels of education.

The areas with the cheapest rents, e.g. Mount Pleasant East and Knight, are in a block in Vancouver East. These may well be an area of low incomes and poorer-quality housing. This is supported by the fact that Mount Pleasant East has the lowest income in all of the city. This may well be a rundown area.

The highest incomes are found in Point Grey and Kerrisdale. These are both in West Vancouver. It is interesting to note that the highest rents (in Coal Harbour and Yale Town) are both in coastal locations, suggesting nice views over the sea or harbour.

6/6 marks awarded Such data response questions with unfamiliar data are difficult to answer and it is often not clear where to start. This student has made a good start by quickly manipulating the data to calculate an average value for city age. Once this has been done, the extremes of age can be linked to the other factors in Table 1, and some valid points are made. Note that some valid explanations are given which cannot be credited, as this question assesses A03 only.

The second and third paragraphs make connections within the data in Table 1 and with Figure 6 – again, a good thing to do. Once again, valid commentary is provided but note that credit can only be awarded for the links made between the data formats.

Question 3

Figure 7 shows information about air pollution in Beijing, China. Using Figure 7, and your own knowledge, evaluate how air pollution in cities affects the people who live in them.

(9 marks)

Daily average air quality index (AQI) at US Embassy, based on PM$_{2.5}$ concentration readings

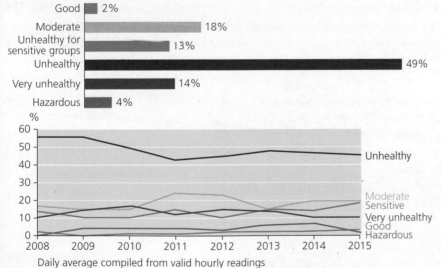

Daily average compiled from valid hourly readings
*AQI categories as set by the US Environmental Protection Agency

Figure 7 Air pollution in Beijing, China

Level 3 (7–9 marks)

AO1 – Demonstrates detailed knowledge and understanding of concepts, processes, interactions and change. These underpin the response throughout.

AO2 – Applies knowledge and understanding appropriately with detail. Connections and relationships between different aspects of study are fully developed with complete relevance. Evaluation is detailed and well supported with appropriate evidence.

Level 2 (4–6 marks)

AO1 – Demonstrates clear knowledge and understanding of concepts, processes, interactions and change. These are mostly relevant though there may be some minor inaccuracy.

AO2 – Applies clear knowledge and understanding appropriately. Connections and relationships between different aspects of study are evident with some relevance. Evaluation is evident and supported with clear and appropriate evidence.

Level 1 (1–3 marks)

AO1 – Demonstrates basic knowledge and understanding of concepts, processes, interactions and change. This offers limited relevance with inaccuracy.

AO2 – Applies limited knowledge and understanding. Connections and relationships between different aspects of study are basic with limited relevance. Evaluation is basic and supported with limited appropriate evidence.

Student answer

Poor air quality is a direct consequence of atmospheric pollution in urban areas. The amount of air pollution depends on the rate at which pollutants are produced and the rate at which they are dispersed (diluted) as they move away from their source. The four key atmospheric pollutants which are likely to have an impact on health are: ozone (O_3), nitrogen dioxide (NO_2), sulphur dioxide (SO_2) and particulate matter. A significant local source of air pollution is traffic emissions. Photochemical smog is another problem – it causes health problems (headaches, eye irritation, coughs and chest pains).

9/9 marks awarded The first paragraph illustrates good knowledge and understanding (AO1) of the topic, giving detail of the polluting gases responsible and the causes of them.

Cities in the developing world are suffering from severe levels of atmospheric pollution, such as China. The Chinese capital, Beijing, has for many years suffered from serious air pollution as shown in Figure 7, with almost half of the daily averages of the Air Quality Index (AQI) being 'unhealthy' according to the criteria of the American Environmental Protection Agency (EPA) based on concentrations of PM2.5 particles. Such particles are small enough to be inhaled and can damage lungs. A further 18% daily averages are classed as very unhealthy and hazardous, making a total of 67%, two-thirds of all the days.

The second paragraph makes use of the data to reinforce the point made earlier and in doing so processes the data (albeit in a simple way) rather than rewriting it.

The main sources of pollutants include exhaust emission from Beijing's five million motor vehicles, coal burning in neighbouring regions, dust storms from the north and local construction dust. From a case study of Beijing I did, a severe smog engulfed the city for several weeks in January 2013, raising public awareness to unprecedented levels and prompting the government to roll out emergency measures. The smog was so bad that air quality readings from the American Embassy said simply: 'Beyond Index'. A reading above 100 is considered 'unhealthy for sensitive groups' and over 400 is rated 'hazardous' for all. At one point a reading of 886 was recorded. Such a reading is hugely dangerous for everybody.

So it is fair to say that air pollution in cities does severely affect the people who live in them.

The third paragraph introduces some case study information which conveniently matches the data – if you have studied a particular case study which features in the data, then you should use it.

Throughout the answer, the student makes the necessary link to impact on people, thereby accessing the AO2 element of the task. The final sentence gives a brief statement of evaluation that summarises what has been written before.

Question 4

Assess the importance of economic change in explaining why some urban areas need to be regenerated.

(9 marks)

Level 3 (7–9 marks)

AO1 – Demonstrates detailed knowledge and understanding of concepts, processes, interactions and change. These underpin the response throughout.

AO2 – Applies knowledge and understanding appropriately with detail. Connections and relationships between different aspects of study are fully developed with complete relevance. Evaluation is detailed and well supported with appropriate evidence.

Level 2 (4–6 marks)

AO1 – Demonstrates clear knowledge and understanding of concepts, processes, interactions and change. These are mostly relevant though there may be some minor inaccuracy.

AO2 – Applies clear knowledge and understanding appropriately. Connections and relationships between different aspects of study are evident with some relevance. Evaluation is evident and supported with clear and appropriate evidence.

Level 1 (1–3 marks)

AO1 – Demonstrates basic knowledge and understanding of concepts, processes, interactions and change. This offers limited relevance with inaccuracy.

AO2 – Applies limited knowledge and understanding. Connections and relationships between different aspects of study are basic with limited relevance. Evaluation is basic and supported with limited appropriate evidence.

Student answer

Economic change is one process that sometimes causes a need for regeneration. The economic structure of an area refers to its balance of primary, secondary and tertiary employment. In many UK and USA cities, economic change has meant deindustrialisation and a loss of jobs in manufacturing and distribution. In the case of Middlesbrough in the UK, this meant a loss of jobs in steelmaking and petrochemicals, whereas in Detroit (USA) the decline of car manufacturing has caused mass unemployment among middle-aged male workers in particular. This leads to a spiral of decline or a negative multiplier effect whereby the loss of jobs in one sector leads to losses in others – such as supplier factories and local services.

9/9 marks awarded The answer begins with a clear definition of economic structure, which demonstrates an understanding of economic sectors. Examples are used to illustrate economic change and the use of another concept, the negative multiplier effect, shows deeper understanding and provides an explanation of why some cities need to regenerate.

However, the consequences of economic change are perhaps more significant, especially population decline. Detroit's population has declined from 1.5 million in 1970 to 0.7 million by 2015. People have left the city because of the loss of jobs, and the lack of new service sector jobs to replace those lost. This is significant because the people who leave tend to be the young and skilled, leaving older less-skilled people behind. Consequently, Detroit is in great need of regeneration to halt further economic decline and outward migration.

Economic change is then linked to further changes in population and the image of a place, exemplified by Detroit, USA, which shows that changes in employment are not the only factor involved. This is assessment, as the answer considers the importance of other factors.

However, regeneration is made less likely in cities that gain a reputation for industrial decline and other urban problems like homelessness and crime. This is the case in Liverpool and Detroit and the negative images of these places then deter investment. This shows that although economic change is a root cause of the need to regenerate, it is made worse by demographic and image changes. The shift towards a service sector economy in the USA and UK since 1970 has left many northern industrial cities isolated from the core area of economic growth, i.e. London in the UK, and in the USA, the core areas are the East and West Coasts. This isolation makes it hard to attract investment, which tends to go to these core areas.

The third paragraph refers to some cities being isolated from the economic core in countries (the UK and the USA) that are now dominated by the service sector – a further factor to be considered.

In conclusion, it can be said that economic change is indeed a key reason why regeneration is needed in many declining urban areas in the world.

The final sentence gives a brief statement of assessment that summarises what has gone before – a necessary concluding statement.

Question 5

In the context of urban areas, to what extent is the management of socioeconomic issues more important than dealing with environmental challenges?

(20 marks)

See the generic mark scheme on page 138.

Student answer

For many people, socioeconomic issues that concern people, their incomes and wellbeing are very important. Issues such as deprivation in urban areas, segregation of ethnic groups, poverty and inequality have a direct impact on the day-to-day lives of people. Urban environmental issues such as air pollution and waste management are perhaps less important – but it does depend heavily on the type of urban area people live in.

In London, environmental issues, especially air pollution, have moved up the political agenda. 95% of areas in London breach WHO guidelines for $PM_{2.5}$ particles. The introduction of the Congestion Charge (2003) and the Low Emissions Zone (2008) are both attempts to manage this issue. These have met with limited success as London's population continues to grow. For many Londoners, social issues such as the affordability of houses are much more important. In 2017, average London house prices were 14 times average London earnings, meaning many people will struggle to ever buy a house.

In Middlesbrough, a deindustrialised town in the northeast of England, the key issue is deprivation. 48% of the population of Middlesbrough ranks in the bottom 10% of most deprived areas in England and unemployment is over 5%. Health and education outcomes tend to be worse than in London. While huge areas of the capital city such as Docklands and Stratford have been improved environmentally and economically by urban regeneration, this has not happened to the same extent in Middlesbrough, where many ex-industrial sites are derelict and contaminated.

The place-context of an urban area has a large impact on which issues people perceive as the most important. It could be said that in wealthy urban areas people have the luxury of being concerned about the quality of green spaces and air quality, whereas in low-income cities, the quality of services, housing and job availability are more important.

In developing world mega-cities such as Karachi (Pakistan), the challenges that many people face are environmental health issues. These include lack of access to sanitation (e.g. open sewers in the streets of Orangi Town, Karachi's largest slum with 2.4 million people) and a lack of access to affordable clean water. Water is often three or four times more expensive bought from a street vendor than from a piped supply, which slum dwellers do not have access to. These issues directly affect quality of life through ill-health, lost working days and high medical costs.

20/20 marks awarded This is a complex question that can be answered successfully in many different ways. This answer is a very good one – its strength is in the use of place-based examples (London, Middlesbrough and Karachi) to support the overall argument that the most important urban management challenges vary depending on the context of the city being considered.

The answer begins with an introduction that states at the outset socioeconomic issues may tend to be more important than environmental issues – providing a sense of how the argument will be constructed.

Two detailed paragraphs then follow that address both issues in the question, with some good comparative statements being made.

These are followed by an excellent evaluative paragraph.

The challenges in Karachi are very different from those in UK cities and, in addition, blur the boundary between socioeconomic and environmental issues. Many areas of Orangi Town have installed their own sewers, usually with the help of NGOs, like the Orangi Town Pilot Project. This type of self-help scheme to overcome challenges is quite common in developing world cities but rare in developed ones, where urban governments are expected to manage their issues.

Overall, the importance of socioeconomic issues versus environmental ones depends on the specific urban area. In low-income areas and cities, socioeconomic needs and high levels of deprivation generally mean the most important focus is on basic needs: decent housing, basic services and incomes people can live on. In richer cities, environmental challenges are more significant because few people lack the basics.

The fifth and sixth paragraphs examine a developing world city – Karachi. Once again, these paragraphs are both detailed and evaluative with good comparative statements being made. The student also argues an interesting and sophisticated point that separating environmental from socioeconomic issues may not be possible in cities such as Karachi as the problems are closely interlinked.

The conclusion states that, overall, socioeconomic challenges are more significant in poorer places and environmental challenges are further up the management and political agendas in wealthier places.

Population and the environment

Question 1

Explain two factors that might account for the variation in immigration around the world.   (4 marks)

1 mark per valid point.

Student answer

The economic growth of countries and their affluence is a significant factor ✓. For example, a wealthy country such as Germany has a higher number of migrants at 450,000 (2015) compared to 1,200 in Romania (2015) ✓. The standards of healthcare and services in more advanced countries will also attract more migrants as the appeal of medicine and access to cheap healthcare of a higher quality will attract more migrants ✓. For example, people from the poorer EU countries (such as Bulgaria), that may not offer free healthcare like the UK NHS system, are attracted by a better quality of life than in their country ✓.

4/4 marks awarded The student provides four correct statements. Note the second and fourth points are development points of the first and third points.

Question 2

Table 2 shows some aspects of development and population for selected countries.
Analyse Table 2. (6 marks)

Table 2 Human development index (HDI) (2019) and Net migration (2018)

Country	Norway	USA	France	Czechia	Iraq	Egypt	India	Nigeria	Malawi
HDI	0.95	0.92	0.89	0.89	0.69	0.70	0.65	0.53	0.49
Net migration	4	3	1	3	2	–1	0	0	–1

Note: net migration = net immigration – net emigration per 1,000 people

Level 2 (4–6 marks)

AO3 – Clear analysis of the quantitative evidence provided, which makes appropriate use of evidence in support. Clear connection(s) between different aspects of the evidence.

Level 1 (1–3 marks)

AO3 – Basic analysis of the quantitative evidence provided, which makes limited use of evidence in support. Basic connection(s) between different aspects of the evidence.

Student answer

The Human Development Index (HDI) includes GDP per capita among the criteria that are used to create it – the higher the HDI, the better the standard of living (and GDP) for the people living in that country. It is clear that richer economies will be more attractive to migrants as they will want to have a higher standard of living, both economic migrants and refugees fleeing persecution. This is shown by the data in Table 2, where all of the four countries with HDI over 0.8 have net migration into them. The relationship isn't that clear, though, with France having a relatively low net migration rate. On the other hand, the countries with lower HDI scores, such as Malawi and Egypt have net out-migration, with a score of minus 1 each.

Two anomalies to this pattern are Iraq and Czechia, which both have a higher amount of net inward migrants than might be expected. The former has a relatively low HDI, yet high inward migration, probably due to being located near to Afghanistan and migrants fleeing the war there may have crossed over the border. The country of Czechia is also an interesting anomaly in that it has a relatively high HDI – the same as France – yet three times as many net in-migrants.

6/6 marks awarded The student recognises that the data have some complexity and although some connections can be seen, they are not all straightforward. Basis relationships within the data are summarised well in the first paragraph, with a clear anomaly also offered (France).

He/She then moves on to examine a number of other anomalous situations, and seeks to explain them. Note that although this explanatory statement may be valid, credit cannot be awarded for it here. The final sentence is another good one. There is good use of both qualitative and quantitative statements throughout this answer.

Question 3

The following extract gives an account of plastic pollution in the world's oceans. Using the extract, and your own knowledge, evaluate how the issue of plastic pollution illustrates the Population, Resources and Pollution model. (9 marks)

Can we tackle the ocean plastics problem?

Most of the plastic debris floating in the oceans is brought there by ten of the world's largest rivers, including the Mekong, Yangtze, Indus and Brahmaputra in Asia, and the Nile and Niger in Africa. These river basins include some of the world's most populous areas. The Yangtze River (1,808,500 km^2) alone supplies about 1.5 million tonnes of plastic debris to the Yellow Sea each year.

Ocean currents transport these riverine inputs great distances and have produced five gigantic zones of floating plastic in the sub-tropical gyres. Who should take responsibility for cleaning up these parts of the global commons? Or should we focus efforts on tackling the problem at source?

A recent initiative based on technology supported by the Dutch government aims to clean up 50% of the large plastic debris floating in the 'Great Pacific garbage patch' that lies in the sub-tropical ocean gyre between California and Hawaii. This is the largest concentration of ocean plastic in the world.

In September 2018, the Ocean Cleanup campaign was launched off the coast of California. A giant floating tube, 600 metres long with a 3-metre screen suspended below, will drift with the ocean currents in a horseshoe shape and gather large plastic debris lying in its path at or near the ocean surface. A support vessel will retrieve the collected plastic for recycling in the Netherlands. The aim is to collect as much of the large floating plastic debris as possible before it breaks down into microplastic particles. The apparatus has been designed to avoid trapping marine mammals and fish.

Level 3 (7–9 marks)

AO1 – Demonstrates detailed knowledge and understanding of concepts, processes, interactions and change. These underpin the response throughout.

AO2 – Applies knowledge and understanding appropriately with detail. Connections and relationships between different aspects of study are fully developed with complete relevance. Evaluation is detailed and well supported with appropriate evidence.

Level 2 (4–6 marks)

AO1 – Demonstrates clear knowledge and understanding of concepts, processes, interactions and change. These are mostly relevant though there may be some minor inaccuracy.

AO2 – Applies clear knowledge and understanding appropriately. Connections and relationships between different aspects of study are evident with some relevance. Evaluation is evident and supported with clear and appropriate evidence.

Level 1 (1–3 marks)

AO1 – Demonstrates basic knowledge and understanding of concepts, processes, interactions and change. This offers limited relevance with inaccuracy.

AO2 – Applies limited knowledge and understanding. Connections and relationships between different aspects of study are basic with limited relevance. Evaluation is basic and supported with limited appropriate evidence.

Student answer

The Population, Resources and Pollution model (PRP) states that people acquire resources from the environment to enhance economic development but such use can result in pollution which damages the same environment the resources have been extracted from. Various feedback loops occur within the three main components, which either enhance the relationships or seek to reduce them.

The ocean plastic issue illustrates that people take resources from the environment to produce the plastic, but people then dispose of that plastic and therefore pollute the environment. A difference from the model is that the resources come from the ground to produce the plastic, in the form of oil, but the pollution is occurring in rivers and the oceans.

A positive feedback loop is that as more plastic is produced to meet the needs of an increasing population, then as living standards increase, more people will demand greater use of resources, which will produce more plastic and even more pollution. Somehow this cycle has to be broken if we want to preserve the oceans.

A negative feedback loop is illustrated in the extract. Here an initiative to collect the plastic in the Pacific Ocean will hopefully reduce the amount of plastic floating in the ocean, so that it can be recycled (in the Netherlands), and thereby reduce the amount of resources being taken from the ground by the rising population.

The PRP illustrates the complex relationship between the three components. Plastic is a resource that is needed by a population, but one which causes pollution if not disposed of properly.

9/9 marks awarded The introductory paragraph illustrates good knowledge and understanding of the PRP model (AO1).

In each of the subsequent paragraphs, the student seeks to make a connection between the PRP and the issue of ocean plastic pollution. The second paragraph makes a simple point, but nevertheless a valid one.

The next two paragraphs are more sophisticated by addressing feedback loops. The fourth paragraph makes use of the information provided in the extract, which is a key requirement of the question. So, throughout the answer the student makes the necessary link to the PRP, thereby accessing the AO2 element of the task.

The final paragraph summarises what has gone before, but nevertheless such a statement is necessary.

Question 4

Assess how the age–sex composition of a population changes at different stages of the demographic transition.

(9 marks)

Level 3 (7–9 marks)

AO1 – Demonstrates detailed knowledge and understanding of concepts, processes, interactions and change. These underpin the response throughout.

AO2 – Applies knowledge and understanding appropriately with detail. Connections and relationships between different aspects of study are fully developed with complete relevance. Assessment is detailed and well supported with appropriate evidence.

Level 2 (4–6 marks)

AO1 – Demonstrates clear knowledge and understanding of concepts, processes, interactions and change. These are mostly relevant though there may be some minor inaccuracy.

AO2 – Applies clear knowledge and understanding appropriately. Connections and relationships between different aspects of study are evident with some relevance. Assessment is evident and supported with clear and appropriate evidence.

Level 1 (1–3 marks)

AO1 – Demonstrates basic knowledge and understanding of concepts, processes, interactions and change. This offers limited relevance with inaccuracy.

AO2 – Applies limited knowledge and understanding. Connections and relationships between different aspects of study are basic with limited relevance. Assessment is basic and supported with limited appropriate evidence.

Student answer

Different population structures of age–sex compositions at different stages of the Demographic Transition Model (DTM) can be shown effectively through the diagrams known as population pyramids. For stage 1 of the DTM, there are high and fluctuating birth and death rates. This gives us a wide base on the population pyramid indicating a youthful population, for example as in Afghanistan. Also, the sharply indenting sides show the high death rate and low life expectancy as there is poor healthcare and technological advances in fighting diseases are non-existent, resulting in few older people.

In stage 2 of the DTM death rates fall while birth rate stays the same. The pyramid shows that there is now a slightly larger elderly population than before because people now live longer – longer life expectancy as healthcare improves but mostly because there is better quality nutrition and better water sanitation systems. Several of the countries of sub-Saharan Africa are thought to fit this pattern, such as Mali and Niger.

7/9 marks awarded This answer examines each of the DTM stages in turn and in each case the analysis is detailed and sophisticated. The first paragraph describes the shape of the pyramid with explanation of the processes behind it, together with an example.

As a country progresses to stage 3 of the DTM, the upper part of the pyramid begins to properly fill out – to widen. There are now much more equal proportions (but not yet equal) between younger and older elements of the population. A larger proportion of middle-aged people between 25 and 45 become apparent, as well as more elderly people. Countries currently at this stage are Brazil, India and Mexico. This 'window' of low economic dependency within the population has been called a 'demographic dividend' but the countries need to have the right level of governance to exploit it.

In stage 4, we see a very balanced pyramid with largely vertical sides because both death rates and birth rates are very low and almost equal. Life expectancy is also higher than before so there are more older people. The UK was in this category until recently when an increase in birth rates caused by migration has produced a small spike in the base of the pyramid.

Lastly, stage 5 is believed to be where some countries, such as Germany, Italy and Japan are entering. This is because birth rates could further drop and we would end up with an elderly or ageing population where the majority are over 60. This is the only place where there may be a sex/gender variation, as women tend to live longer than men, so there are more elderly women in this stage.

This strategy is continued in the next two paragraphs, with the third paragraph being particularly sophisticated with a reference to the demographic dividend and governance.

The final two paragraphs complete the task with Stages 4 and 5 of the DTM, with a neat little reference to gender differences at the end. Overall this a very thorough answer, but lacks a summative sentence explicitly addressing the 'assessment' element of the question, even though evaluation has been provided implicitly throughout.

Question 5

'Infectious diseases have a greater impact on economic development than non-communicable diseases'. To what extent do you agree with this view?

(20 marks)

See the generic mark scheme on page 138.

Student answer

Many would agree with this statement, because infectious diseases can be spread easily from person to person and so if they are not controlled, they can affect large proportions of a population quickly and therefore have a great impact on economic development. To examine this issue in more detail, this essay will concentrate on two diseases – malaria, an infectious disease, and cancer, a non-communicable disease (NCD).

Malaria has serious impacts, slowing economic growth and development, and prolonging the vicious cycle of poverty that exists in many developing countries, especially in sub-Saharan Africa. There are several costs involved both to individuals and governments. Costs to individuals and their families include the purchase of drugs for treating malaria, expenses for travel to (and treatment at) clinics, lost days of work, reduction in crop production, and expenses for preventative measures and ultimately burial after death. Costs to

18/20 marks awarded This is a coherent, logical and sophisticated answer. It begins with a clear introduction, and the student clearly states that he/she will concentrate on two identified diseases.

governments include the supply and staffing of health facilities, the purchase of drugs, public health interventions such as insecticide spraying or distribution of insecticide-treated bed nets and lost opportunities for tourism.

It has been estimated that in some parts of sub-Saharan Africa, malaria accounts for up to 40% of public health expenditure. Direct costs (for example, illness, treatment and premature death) have been estimated to be at least US$12 billion per year. The cost in lost economic growth is many times more than that, and is probably impossible to measure. So, for many poorer countries, infectious diseases such as malaria have severe economic impacts because they do not have the resources to effectively manage the disease.

The subsequent paragraphs provide a sophisticated summary of the economic impacts of malaria on individuals and governments, which ends with some impressive statistics which seem plausible, though perhaps impossible to verify.

Cancer is an NCD – it cannot be caught by transmission. It is expected to kill over 20 million a year by 2030. Its rise is thought to be caused as a result of the misuse of alcohol and tobacco, and poor diet (not enough fruit and vegetables). It has also been shown that there is a strong correlation between age and being susceptible to cancer. Cancer can prevent you from working, as treatment is often long, and involves the use of drugs and radiology. This means that you will suffer a loss of income. This loss of working time will have a negative effect on a national economy. Also, it is likely that sufferers of cancer will be looked after by relatives who are of working age, so this also means they will be losing an income and affecting the economy.

The answer then considers the impact of cancer in general. Perhaps it might have been better to concentrate on one form of the condition, but that might not have been the decision of the student – the teacher could have chosen to study 'cancer' in general. Although not quite at the same level or as detailed as the discussion on malaria, some interesting and valid points are made.

Comparing the two types of disease, NCDs are more likely to occur in later life. On the other hand, infectious diseases can affect anyone, and children can die from them. This means that infectious diseases cause a greater loss of the workforce, as high infant mortality rates mean that children will not grow up and join the workforce. Therefore, infectious diseases slow economic growth in the longer term more than NCDs. On the other hand, people with an infectious disease may recover quickly if they are treated and so the overall loss to productivity could be lower.

The student then begins to compare the economic impact of the two types of disease considered and some good discursive points are made.

The effect on economic development also depends on the country studied. In richer countries, such as the UK, most infectious diseases are treated easily and are prevented by vaccinations. NCDs are more common but healthcare is available to treat the symptoms. In contrast to this, a poorer country may suffer more from an infectious disease and may increasingly suffer from NCDs too due to changing lifestyles as they become more westernised. Therefore I think that an overall assessment of the quotation in the question is more complex – for a developed country an NCD has the greatest impact on economic development and for a developing country an infectious disease has the greatest impact.

The concluding paragraph continues an evaluative approach and a clear overall view with regard to the statement in the question is provided.

Resource security

Question 1

In the context of resources, explain the difference between measured reserves and indicated reserves. (4 marks)

> 1 mark per valid point.

Student answer

Measured reserves exist where the quantity and quality of a resource are known such that their worth can be estimated with a high level of confidence ✓. They are viewed as being economically viable ✓. On the other hand, indicated reserves are where the quantity and quality of a resource are only well-known enough that they need to be further evaluated as to their economic viability ✓. Confidence in measured reserves is therefore stronger ✓.

4/4 marks awarded The student provides four correct statements.

Question 2

Study Figure 8, which shows UK electricity generation by renewable sources (2000–2017). Analyse the information provided in Figure 8. (6 marks)

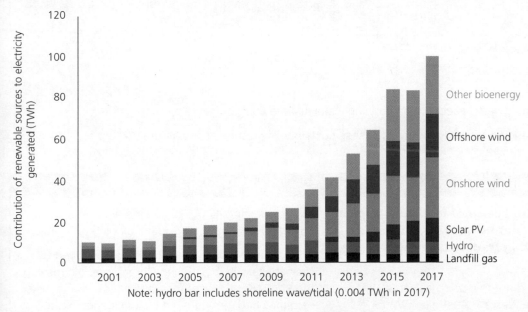

Note: hydro bar includes shoreline wave/tidal (0.004 TWh in 2017)

Figure 8 UK electricity generation by renewable sources (2000–2017)

Source: ONS (2018)

Student answer

Electricity generation in the UK from renewable sources has increased greatly this century from 10 TWh in 2000 to 100 TWh in 2017 – a tenfold increase. The increase has been consistent throughout this period, with only a few years of minimal growth, e.g. 2000 to 2003 and 2015 to 2016 (a slight decline). Some sources such as landfill gas and HEP have shown modest increases (even a small decline recently) whereas others such as wind power (both offshore and onshore) have grown much more rapidly. Wind power, from both locations, now accounts for about 60% of all such energy. 'Other bioenergy' has also increased greatly over the time period and is now at about 25% of the total. The use of solar PV has developed and increased only within the last 6 years of the data, reaching about 15% of the total. This energy is from solar panels on house roofs – and it has the potential to increase further if government subsidies are continued.

5/6 marks awarded The student addresses the question and makes a number of clear quantitative and qualitative statements. However, the answer is a bit of a 'Cook's Tour' of the data, and, other than the statement regarding wind power, lacks a degree of overview. Level 2 criteria have been met, and so mid-Level 2 credit can be awarded.

Question 3

The following extract gives an account of a mining issue in June 2020.

Using the extract, and your own knowledge, assess sustainability issues associated with mineral ore extraction.

(9 marks)

Western Australia

Mining giant BHP vows not to disturb ancient Aboriginal sites dating back 15,000 years without consulting traditional owners of the land

BHP has promised not to disturb ancient Aboriginal sites in Western Australia (WA) for a mine expansion without consulting traditional owners, following outrage over Rio Tinto blowing up 46,000-year-old rock shelters in the same region. BHP applied in October to work in an area containing 40 significant indigenous sites to enlarge the $4.5 billion South Flank iron ore project.

That came five days after Rio Tinto destroyed two rock shelters in the Juukan Gorge area, distressing the Puutu Kunti Kurrama and Pinikura people, who said they had mentioned the significance of the sites 'for years'. One of the sites BHP applied to develop is in the Djadjiling Range, which is also in the Pilbara and contains rock shelters and has been occupied for a similar amount of time.

A minister in the Australian government has stated that he wants impacts to Aboriginal sites 'limited to the practical extent possible' but believes in indigenous self-determination and supports native title groups using their hard-won rights to make commercial deals.

A BHP spokeswoman said the company had consulted extensively with the Banjima people over 15 years and would continue to do so. 'We take a sustainable approach to our mining operations and work in partnership with traditional owners to ensure that each stage of development is informed by their views,' she said. 'We will not disturb the sites identified without further extensive consultation with the Banjima people. This will include further scientific study and discussion on mitigation and preservation.'

The WA government hopes to pass its new Aboriginal cultural heritage bill this year, which will provide for agreements between traditional owners and companies to consider new information and be amended by mutual consent.

Level 3 (7–9 marks)

AO1 – Demonstrates detailed knowledge and understanding of concepts, processes, interactions and change. These underpin the response throughout.

AO2 – Applies knowledge and understanding appropriately with detail. Connections and relationships between different aspects of study are fully developed with complete relevance. Evaluation is detailed and well supported with appropriate evidence.

Level 2 (4–6 marks)

AO1 – Demonstrates clear knowledge and understanding of concepts, processes, interactions and change. These are mostly relevant though there may be some minor inaccuracy.

AO2 – Applies clear knowledge and understanding appropriately. Connections and relationships between different aspects of study are evident with some relevance. Evaluation is evident and supported with clear and appropriate evidence.

Level 1 (1–3 marks)

AO1 – Demonstrates basic knowledge and understanding of concepts, processes, interactions and change. This offers limited relevance with inaccuracy.

AO2 – Applies limited knowledge and understanding. Connections and relationships between different aspects of study are basic with limited relevance. Evaluation is basic and supported with limited appropriate evidence.

Student answer

Mineral ore extraction can create a number of sustainability issues. Many of the issues are connected with pollution, of land, air and water, economic viability and cultural destruction against the wishes of indigenous people, some of which are highlighted in the extract.

Pollution can be visual, with noise, or atmospheric. It can also be extremely toxic if it enters water sources. In the Bento Rodrigues event in Brazil in 2015, the sludge that was released by the dam collapse not only destroyed houses and buried people, it affected drinking water, animals and fish in the rivers and even animals (such as the leatherback turtle) in the seas off the coast of Brazil as the slurry entered the sea. The dirty water in the rivers can also block sunlight for aquatic plants and cause them to choke.

8/9 marks awarded Note that this synoptic question connects the content areas of resource extraction, place-meaning and sustainability. This introduction recognises that there are a number of sustainability issues, and that they are not all environmental. They can also include economic sustainability and cultural sustainability.

The above represents a form of environmental sustainability, but other aspects such as economic sustainability should be considered too. The Bento Rodrigues event will cost a lot of money to clean up – money which could have been used for other purposes such as education and health in such an emerging country. It is interesting to note that the Brazilian government is seeking compensation from the TNCs that owned the iron ore mine. Local fishermen are also seeking compensation.

Finally, another sustainability issue is that the culture of indigenous people can be damaged if the necessary permissions are not obtained. This is the issue in the extract, where two large TNCs (BHP and Rio Tinto) are at odds with the local aboriginal population – Rio Tinto certainly overstepped the mark when they blew up some ancient rock shelters. The problem is one of balancing economic development which would benefit the aboriginal people of the area, but this can upset others in the same community who value cultural features as being more important.

So, overall, the information provided in the extract does highlight a number of sustainability issues for which the solutions are not straightforward.

Political factors may also play a role. The answer addresses these separately in the second, third and fourth paragraphs, and makes several references to the data in support, but crucially does not depend on it. It is important to note that the answer refers to the data provided, as required, but also refers to other case studies that have been covered.

This is a conceptually strong answer, and is well argued. It ends with a brief concluding statement, which is always a good thing to do.

Question 4

Evaluate the extent to which conflicts can arise over water security.

(9 marks)

Level 3 (7–9 marks)

AO1 – Demonstrates detailed knowledge and understanding of concepts, processes, interactions and change. These underpin the response throughout.

AO2 – Applies knowledge and understanding appropriately with detail. Connections and relationships between different aspects of study are fully developed with complete relevance. Evaluation is detailed and well supported with appropriate evidence.

Level 2 (4–6 marks)

AO1 – Demonstrates clear knowledge and understanding of concepts, processes, interactions and change. These are mostly relevant though there may be some minor inaccuracy.

AO2 – Applies clear knowledge and understanding appropriately. Connections and relationships between different aspects of study are evident with some relevance. Evaluation is evident and supported with clear and appropriate evidence.

Level 1 (1–3 marks)

AO1 – Demonstrates basic knowledge and understanding of concepts, processes, interactions and change. This offers limited relevance with inaccuracy.

AO2 – Applies limited knowledge and understanding. Connections and relationships between different aspects of study are basic with limited relevance. Evaluation is basic and supported with limited appropriate evidence.

Student answer

Conflicts over water security can take place where there are trans-boundary water sources. Many trans-boundary water sources straddle an international boundary such as the River Nile (Egypt, Ethiopia and Sudan). Where conflict does exist, it ranges on a spectrum from mild, diplomatic disagreement to the very rare situation when water becomes a source of war.

Water is a precious resource in Egypt and Sudan. Cairo in Egypt receives only 25 mm of rainfall per year and the country depends almost entirely on the River Nile for its water supply. Historic agreements from the colonial era gave Egypt and Sudan rights to all of the Nile's waters. Upstream countries signed the Nile Basin Initiative in 1999 but Egypt and Sudan refused to be involved. Increased water usage upstream, and the prospect of a huge new HEP dam in Ethiopia, risk reducing necessary Nile River flows reaching Egypt and Sudan.

A similar situation exists on the Mekong River in Asia. Upstream dam construction by China risks the water supply to downstream Vietnam, Laos and Cambodia. The latter countries are part of a water-sharing treaty called the Mekong River Commission, but China is not. In both the Mekong and Nile cases, the long-term solution is for all trans-boundary basin countries to enter into an agreement based on existing global rules.

Trans-boundary water supplies have only rarely contributed to armed conflict. Between 1964 and 1967 there were a series of military clashes between Israel and its Arab neighbours (Syria, Palestine) over control of the River Jordan. Even this conflict has had some resolution, with Israel and Jordan signing a water-sharing agreement in 1994.

In conclusion, it is not the case that trans-boundary water supplies always lead to conflict. There are more examples of water-sharing agreements than conflicts, and recognised international frameworks for resolving disputes. Where conflict does exist, it often exists as part of wider political disputes.

9/9 marks awarded This is a well-supported answer that applies a number of examples and case studies effectively to answer the question. The definition of 'trans-boundary' is a good way to start, as it focuses the answer on the key topic of the question. Defining conflict is also sensible, as it shows understanding of the concept, i.e. it is much more than simply 'war'.

The level of detail provided on the situation regarding the River Nile is good, and the use of the Mekong example supports the argument made about the Nile, i.e. that conflict exists when all parties are not part of an agreement.

The River Jordan example makes the useful point that armed conflict is very much the exception not the norm.

The final statement, that trans-boundary water conflicts often exist when there are pre-existing unrelated political issues, recognises the complexity of the situation. Writing a brief conclusion to such a question is good examination technique.

Question 5

'Renewable energy sources can meet future global energy demand.' To what extent do you agree with this statement?

(20 marks)

See the generic mark scheme on pages 138.

Student answer

Global energy demand is expected to increase by 40–50% by 2040. Almost all this increased demand is driven by emerging and developing countries. Projections from the US Energy Administration suggest there will be almost no growth in energy consumption in developed countries. This means there are two sub-questions here: can renewable energy replace fossil fuel used in the developed world, and can renewable energy be the major source of future energy supply in other countries?

Globally, 80% of energy production today is from fossil fuels. In developed countries this is changing in two ways. Firstly, there is a switch from coal to natural gas, which is cleaner. Secondly, there is a shift to renewables, especially wind power. Denmark gets 40% of its electricity from wind power and Germany 10%, but globally it only accounts for 4% of electricity. Wind's intermittent and unreliable nature means that standby power stations (usually gas-fired) need to be available to boost supply.

Hydro-electric power (HEP) has been widely developed in some countries. However, it requires specific geographical conditions, i.e. a reliable water supply and valleys that can be flooded to create HEP reservoirs. Canada, Brazil and Norway all generate over 50% of their electricity from HEP. Most developed countries have already utilised suitable sites and have limited capacity to expand. This is not the case in some developing and emerging countries such as Ethiopia and China, where it is rapidly expanding.

Demand in developing and emerging countries is often met by constructing new coal, gas or oil power stations. This is because they are cheap (especially coal), the technology is relatively simple and they can be constructed quickly. In comparison, renewable alternatives have disadvantages. Wind and solar power are more expensive (though costs are falling) and intermittent. HEP has long construction times and frequently involves the displacement of people to create reservoirs. Nuclear power is technically very difficult, and at up to $10 billion for one power station has very high initial costs. Cost is usually the key variable, and this makes it likely that coal and gas will be the most used fuels to meet demand in the future. Coal demand is expected to increase by about 30% to 2035.

20/20 marks awarded This is a strong, well supported answer. From the start, it shows understanding of 'future global energy demand', recognising that future demand will not be the same everywhere, and it is a complex issue. This is turned into a structure for the answer by posing two sub-questions about the developed versus the developing/emerging world.

This is followed by a discussion of wind power, its pros and cons, and the extent to which it has grown in some countries. A similar paragraph on HEP then follows. These paragraphs and the next sections are evaluative because they consider the limitations of wind power and other renewables, especially in terms of their geographic limitations and economic cost.

Cost is identified as the most significant variable in the choice of which energy sources to use, which involves making a judgement, i.e. evaluating.

China shows what can be possible. In 2016, it was the world's largest user of wind power. Capacity increased from 1250 MW in 2005 to 150,000 MW in 2016, but wind still accounted for only 4% of China's total electricity generation.

The detail on wind power in China is a useful case study that adds depth to the answer.

It is questionable whether renewable energy will replace crude oil used to make transport fuels (petrol, diesel) that accounts for 25% of global energy use. Renewable alternatives are not well developed. Biofuels can replace petrol and diesel, but they require large areas of land to grow crops which are increasingly needed to feed a growing world population. Electric vehicles are currently charged by power stations burning fossil fuels. They could be powered by renewably generated electricity but that is a long way off in most countries.

The section on transport fuels makes a strong argument against a renewable future. This type of clear argument is much better than sitting on the fence, and shows the student is prepared to have a view on the topic.

In conclusion, in developed countries renewable energy will increase its share of energy use, but total demand is static because of improving efficiency. In the developing and emerging world, renewables will not meet future demand as long as fossil fuels are a cheaper option. Transport fuel demand is likely to be met by crude oil for the foreseeable future because no viable alternative currently exists.

There is a clear conclusion, which returns to the theme from the introduction that future demand in the developed and developing world need to be considered differently. The answer is evaluative throughout, makes supported judgements and has a clear conclusion.

◼ Questions: Set 2

Global systems and global governance

Question 1

Explain the term 'global marketing' and give an example of it. (4 marks)

> 1 mark per valid point.

Student answer

Global marketing is when a manufacturing or a service company views the world as one market ✓. Its ultimate goal is to sell the same item or provide the service the same way, everywhere ✓. Coca-Cola uses the same formulae (one with sugar, the other with corn syrup) for all its markets ✓. The classic contour bottle design is incorporated in every country, although the size of bottle and can in which it is marketed is the same size as other beverage bottles and cans in any particular country ✓.

4/4 marks awarded The student provides four correct statements. Note that the third and fourth statements are developments of the first and second points respectively.

Question 2

Figure 9 and Table 3 provide information regarding Antarctic tourism from the International Association of Antarctica Tour Operators (IAATO). Analyse Figure 9 and Table 3.

(6 marks)

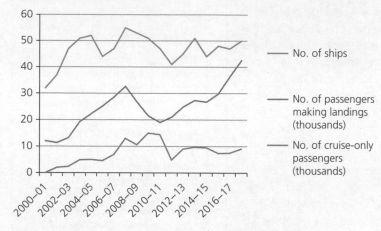

— No. of ships

— No. of passengers making landings (thousands)

— No. of cruise-only passengers (thousands)

Figure 9 Ship-based tourism to Antarctica (2000-2018)

Source: IAATO

Table 3 Antarctic cruise ships: passenger nationalities (2017–18)

Nationality	% of total for 2017–18 season
USA	33.0
China	15.8
Australia	11.2
Germany	7.3
UK	7.0
Canada	4.5
France	4.0
Switzerland	2.0
Netherlands	1.6
Others	13.9

Source: IAATO1

Level 2 (4–6 marks)

AO3 – Clear analysis of the quantitative evidence provided, which makes appropriate use of evidence in support. Clear connection(s) between different aspects of the evidence.

Level 1 (1–3 marks)

AO3 – Basic analysis of the quantitative evidence provided, which makes limited use of evidence in support. Basic connection(s) between different aspects of the evidence.

Student answer

Table 3 shows that 60% of the visitors to Antarctica come from just three countries, USA, China and Australia, with the USA being the most dominant – twice as many as the next country, China. European countries in the table are quite significant too – 22% of the tourists. The great majority of visitors are from developed countries where personal incomes are high – it is not cheap to go to Antarctica on holiday. China is an emerging country and some people in that country are now becoming wealthy.

This is probably a reason for the increase in tourists landing on Antarctica since 2010, as shown in Figure 9. There was a spike earlier on in the century, but this reduced around the time of the world recession in 2007–08. The number of ships increased in the early part of the century, but since then it has remained stable, as too has the number of cruise-only passengers (this has slightly fallen in fact since 2011). This is probably due to the fact that people probably now think that if I am going that far I might as well walk on Antarctica as not – hence the rapid increase in passengers making landings.

6/6 marks awarded The student addresses the data in reverse order and makes some good analytical points in the first paragraph that manipulate the data, e.g. adding up the top three items, 'twice as many', and the total for European countries. The next two sentences go beyond the data given, essentially giving possible reasons. These points cannot be credited here, as this question assesses AO3 only.

However, the possible link between Chinese tourists and an increase in the number of landings at the start of the second paragraph is appropriate – the student is attempting to connect the data. Much of the remainder of the answer analyses the data, though in a rather simple manner. However, in the final sentence the student does try to connect two elements of Figure 9.

Question 3

Figure 10 shows information regarding economic growth in the world, and groups of countries around the world. Using Figure 10, and your own knowledge, assess the view that global economic growth has slowed down. (6 marks)

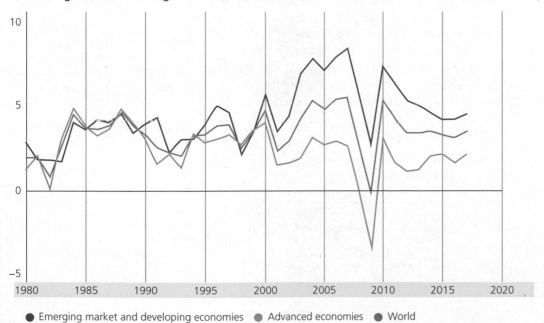

● Emerging market and developing economies ● Advanced economies ● World

Figure 10 Annual percentage change in GDP per capita: world, advanced countries and emerging/developing countries (1980–2017)

Level 2 (4–6 marks)

A01 – Demonstrates clear knowledge and understanding of features, concepts, processes, interactions and change.

A02 – Applies knowledge and understanding to the novel situation, offering clear evaluation and analysis drawn appropriately from the context provided. Connections and relationships between different aspects of study are evident with clear relevance.

Level 1 (1–3 marks)

A01 – Demonstrates basic knowledge and understanding of features, concepts, processes, interactions and change.

A02 – Applies limited knowledge and understanding to the novel situation, offering only basic evaluation and analysis drawn from the context provided. Connections and relationships between different aspects of study are basic with limited relevance.

Student answer

Figure 10 shows that there have been fluctuations in global economic growth since 1980 with a marked decline in 2008–2009 following the global world recession. There was a rapid increase in growth in 2010, but since then there has been further decline, so there clearly has been a slowing down of global economic growth in recent years. It is interesting to note that in the build-up to the recession, the emerging and developing countries were growing fastest, and again after 2009. However, in the last few years, their rate of growth has reduced relative to the developed countries.

6/6 marks awarded The answer gets off to a weak start, in that the student concentrates on describing the data shown in Figure 10. This is not creditworthy until it is linked to the question. However, within the first paragraph, two simple statements are made regarding a slowdown of economic growth.

The global financial crisis from 2007 to 2009 was clearly a major event, and global GDP growth has yet to fully recover. A global slowdown in some international trade has occurred. Some economists have said the period since 2009 can be viewed as a cyclical downturn following the boom that began in the late 1980s. The economic growth rate of the main emerging economy, China, has halved recently, although it is still much larger than many developed world economies. China is deemed by many to be globalisation's driver of economic growth and the slowdown of one of the world's largest economies has serious implications for everyone.

The second paragraph is stronger, demonstrating both knowledge and understanding, and application of that knowledge to address the question.

Added to this is the fact that some barriers to trade are now being talked about and put in place. The USA under President Trump introduced tariffs on goods such as steel, with threats of even more in the future. He threatened specific tariffs on China during the 2020/21 coronavirus crisis. These 'trade wars' will have a negative effect on global economic growth.

The third paragraph also introduces some up-to-date information regarding the USA, which always impresses an examiner. Once again knowledge is demonstrated with some application.

In conclusion, it appears that global economic growth has indeed slowed down and may continue to do so in the years ahead.

The final summative statement addresses the thrust of the question, and pulls together the arguments made mostly in the second and third paragraphs.

Question 4

To what extent can the use of Antarctica be sustainable?   (20 marks)

See the generic mark scheme on page 138.

Student answer

Antarctica has been protected by the Antarctic Treaty System (ATS) since 1961. This recognises the continent as a global commons, a universally owned resource that is managed by international organisations and treaties, rather than individual countries or companies. The question is whether, in an increasingly globalised world, Antarctica can be managed in such a way that it is sustainable, i.e. its resources are conserved so that they are passed on to future generations in the same state they exist in today. Globalisation not only involves increased global trade but also increasing resource consumption and pollution, and increasing flows of people as both migrants and tourists. It is debatable whether Antarctica can be sustainable under these conditions.

On the one hand, the ATS has protected the continent well so far. Economic activities such as mineral exploitation are banned. There are no permanent residents, or facilities for tourism. Only scientific research is permitted. It is unlikely that this will change in the near future. One benefit of globalisation is that Antarctica, and especially its penguin colonies, is globally known via the media (e.g. the television programme *Planet Earth*) and so people are aware of its uniqueness and will want the continent protected.

However, its resources are exploited. About 40,000–50,000 tourists visit each year. Tourists stay on cruise ships, but many land briefly on the continent. Tourism is governed by the regulations of the International Association of Antarctica Tour Operators (IAATO), which are designed to protect the landscape and biodiversity. However, increasing tourism numbers could threaten the sustainability of the ice-free fringes of the continent.

Fish stocks in the Southern Ocean are threatened by overfishing of species such as krill and the Patagonian toothfish. Not enough is known about fish stocks to determine the safe sustainable yield in some cases, and policing illegal fishing is a problem. As globalisation leads to greater affluence and people demand more protein-rich diets, pressure from overfishing is likely to increase. At one time, whaling in the Southern Ocean threatened the survival of many whale species, but it has been successfully turned around by the International Whaling Commission (IWC). Today only a very small number of whales are hunted by Japan (one of only a few countries to contravene the IWC ban on whaling).

20/20 marks awarded The introduction is a substantial one, but one that makes clear that the question is understood, and that the theme is a complex one. The student makes good use of knowledge, as well as introducing an important factor that plays a significant role – globalisation.

The subsequent paragraphs then examine different aspects of the 'use' of Antarctica – resources (physical and natural), tourism and fishing – in each case including a statement regarding sustainability.

apidly increased trade, which has resulted in huge
ntries such as China. However, the environmental
sation is that atmospheric carbon dioxide levels
00ppm. Pollution leading to global warming
widespread impacts on Antarctica's cryosphere and
ecosystems, on land and in the Southern Ocean. There are already
signs of cracks in ice shelves on the Antarctic Peninsula. This is by
far the most significant threat to a sustainable future for this global
commons.

In conclusion, globalisation has made people more aware than ever of
Antarctica's unique position as a global commons. There is evidence
that the continent can be managed sustainably, as can be seen with
tourism. Where there are problems such as overfishing, there are
models of sustainable resource management that could be applied
to Antarctica if international organisations such as the UN and ATS
can muster the political will. However, global governance of pollution
levels to reduce the threat of climate change is required if Antarctica
is to have a long-term, sustainable future.

The penultimate paragraph then returns to the connection with globalisation, focusing on pollution and climate change.

The concluding paragraph draws all of the previous ideas together, and makes the point that whether or not the use of Antarctica can be sustainable, the outcomes are going to be challenging in the future. Bearing in mind the time allowed to answer this question, this answer is excellent.

Changing places

Question 1

Outline how you can investigate a place using qualitative data.

(4 marks)

1 mark per valid point.

Student answer

I can use qualitative data such as photographs or text, from varied
media, audio-visual media, artistic representations and oral sources,
such as interviews, reminiscences, songs etc. ✓ Most people now
have digital cameras on their mobile phones which can use wide
angle and close-up shots. ✓ These can be used to monitor changes
over time. Conducting interviews with people can be useful in finding
out how a person feels about a place ✓, and they can be used to
discover their changing emotions if there has been both continuity
and change in that community ✓.

4/4 marks awarded The student provides four correct statements. Note that the final point is a development of the preceding point.

Question 2

Table 4 shows census data for two places within London – Shadwell and Hampstead Garden Suburb. Analyse the contrasts shown in Table 4. (6 marks)

Table 4 Census data for two London districts (2011)

Census criteria (2011)	Population density per ha.	Population under 17 years of age (%)	Population aged 17–64 (%)	Population aged 65 and over (%)	% of households with an average of more than one person per room	Outright ownership of housing (%)	Households with two or more cars/ vans (%)
Shadwell	399	31.8	64.3	3.9	24.5	3.6	3.8
Hampstead Garden Suburb	11.9	23.9	57.1	19.0	0.6	48.4	57.8

Source: ONS

Level 2 (4–6 marks)

AO3 – Clear analysis of the quantitative evidence provided, which makes appropriate use of evidence in support. Clear connection(s) between different aspects of the evidence.

Level 1 (1–3 marks)

AO3 – Basic analysis of the quantitative evidence provided, which makes limited use of evidence in support. Basic connection(s) between different aspects of the evidence.

Student answer

There is a clear difference in the population of the two areas as the percentage of over 65-year-olds in Hampstead Garden Suburb (HGS) is almost five times higher than that of Shadwell. This will mean that there will be greater provision for the elderly in this area with probably a relatively large amount of care homes and similar residential homes. It also reflects the very low occupancy rate (% of households with an average of more than one person per room), as many elderly people live on their own.

This is also an indicator of overcrowding – there is a significantly greater problem in Shadwell, with nearly a quarter of the households having an average of more than one person per room. This is symptomatic of the fact that the houses will be much larger in HGS, possibly old Victorian properties, and in Shadwell there could be more small terraced houses housing larger families. This is supported by the huge difference in population density – over 30 times greater in Shadwell.

Finally, HGS has a much higher car ownership and home ownership. All this points to HGS being a wealthy area, with many wealthy retired people living in large houses and having two or more cars, and a large proportion of low income people, possibly students or immigrants, living in Shadwell.

6/6 marks awarded The student has provided detailed evidence of the contrasts shown in the data, and has used both quantitative and qualitative statements in doing so. These are sufficient to gain maximum marks. It should also be noted that the student has also been tempted to go beyond the data and comment on the socioeconomic nature of the two areas – in the statements in the second paragraph on types of housing, and in the final sentence of the answer. Although these statements may be valid, they cannot be credited as this question assesses AO3 only.

Question 3

The following extract and Figure 11 provide information about the representation of the Lake District and the town of Keswick. Using the extract and Figure 11, and your own knowledge, assess the ways in which these places have been represented. (6 marks)

Keswick as an adventure holiday hub

The Lake District has been rebranded several times in the past. Until about 300 years ago, it was viewed as barren and wild. It was not until the eighteenth century that natural landscapes were seen as something to be admired. Attitudes towards the Lake District began to change after 1750, when a number of influential travel guidebooks reframed the landscape as picturesque (like a painting) and sublime (awe-inspiring and wonderful). Thomas West's *Guide to the Lakes* (1778) described the rocky scenery in Borrowdale as 'sublimely terrible' and likely to inspire

'reverential awe'. Such comments encouraged visitors to the area, although most admired the scenery from the valley floor, rather than venturing onto the hills.

A further rebranding occurred in 1951, when the area was designated a National Park. In 2017, the Lake District gained World Heritage status from the international body UNESCO. Furthermore, the Lake District, and especially the town of Keswick (see Figure 11), is keen to become recognised as the Adventure Capital of the UK.

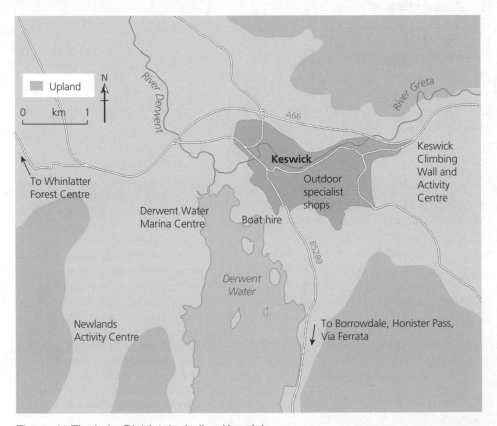

Figure 11 The Lake District, including Keswick

Level 2 (4–6 marks)

AO1 – Demonstrates clear knowledge and understanding of features, concepts, processes, interactions and change.

AO2 – Applies knowledge and understanding to the novel situation, offering clear evaluation and analysis drawn appropriately from the context provided. Connections and relationships between different aspects of study are evident with clear relevance.

Level 1 (1–3 marks)

AO1 – Demonstrates basic knowledge and understanding of features, concepts, processes, interactions and change.

AO2 – Applies limited knowledge and understanding to the novel situation, offering only basic evaluation and analysis drawn from the context provided. Connections and relationships between different aspects of study are basic with limited relevance.

Student answer

The extract and Figure 11 illustrate various ways in which the Lake District and the town of Keswick have been represented in the past in books, and more recently on a map. In all cases they are seeking to capitalise on the region's outstanding natural resources, the landscape, and considerable experience in outdoor recreational provision. Such representation of the area has been important for almost 300 years as guidebooks such as that of Thomas West's in 1778 testify. The emotive words used in the extract – 'picturesque', 'sublime' and 'reverential' – all support this view.

When designation as a National Park took place in 1951, the main purpose was to generate tourist revenue, both directly through employment, for example in outdoor adventure centres, and indirectly, for example when farmers supply local food to hotels and restaurants.

All of the above and the successful bid to become a World Heritage area strived to engender a greater appreciation for, and interest in, conserving the landscape. They also provide facilities that locals as well as visitors can use, thus enriching community life for all people in the area.

The facilities shown in the map of Keswick, Figure 11, all with a view to creating an adventure holiday hub, also fit in with the need to encourage health-related fitness, an important government initiative.

So, a variety of forms of representation have been used, and they all have adventure and exercise as common themes.

6/6 marks awarded The student has engaged with the theme of evaluation throughout. References are clearly made to the stimulus material, but then in most cases the student makes commentary on how these materials support the idea of representation.

The answer ends with a brief, but necessary, conclusion.

Question 4

'Place-shaping processes depend largely on the actions of community groups.' To what extent do you agree with this statement?

(20 marks)

See the generic mark scheme on page 138.

Student answer

Place-shaping is a process aiming to improve the quality and image of areas to make them attractive and useable for existing residents, newcomers and investors. It involves identifying existing problems of places, consulting widely about potential solutions and then developing and implementing plans (such as rebranding) for change.

An example where community groups have been key to place-shaping is the Coin Street redevelopment on London's South Bank near the iconic OXO tower. Coin Street Community Builders (CSCB) is a community social enterprise set up in 1984 that has successfully regenerated and rebranded a once rundown area. The OXO building, four social housing cooperatives and the Bernie Spain Gardens open space have all contributed to a thriving mixed-use development. CSCB aimed to create a mixed community in terms of age, ethnicity and gender in the Coin Street area. There is a wide range of ethnicity in the area, and so CSCB were keen to involve many members of the BME groups in the scheme.

Other players have been important to CSCB's success, including government funding for affordable housing developments and private investment from bars and restaurants such as The Wharf and Studio Six in the commercial spaces. However, the overall vision is a community-led one and their role has been to coordinate the investment of others as well as raise money and plan change themselves.

Coin Street contrasts with more top-down approaches to place-shaping where investors, planners and architects take the lead role such as Salford Quays, Manchester. Landmark buildings in Salford Quays include The Lowry theatre and gallery and the Imperial War Museum North. Nearby MediaCityUK was developed by the private investment Peel Group as a media hub with the BBC, ITV and Salford University among the investors. This type of outward-looking place-shaping and re-imaging has much less input from community groups. In the case of Salford Quays, the old industrial area had few existing residents. The key factors here were commercial investment and decisions by local planning authorities such as Salford City Council.

20/20 marks awarded The answer is logical, with detailed use of appropriate case studies – Coin Street (London), Salford Quays (Manchester) and Brick Lane (London). The initial paragraph sets the scene for what follows, providing an explanation of the meaning of 'place-shaping'.

The next two paragraphs provide precise detail of the Coin Street regeneration area, together with a sense of evaluation that links to the thrust of the question.

These are then followed by a paragraph on Salford Quays that presents an alternative viewpoint about the role of community groups in place-shaping. Again, detail of the area is provided alongside evaluation and with an explicit link to the question.

Community groups have also been responsible for transforming Brick Lane in Tower Hamlets, London. Brick Lane is known for its vibrant Bangladeshi community and their businesses have transformed the area to become the 'curry capital' of the UK. Street artists have also made the area famous for its graffiti. However, the arts, the food and the culturally diverse thriving community have led to increasing accommodation prices. Furthermore, widespread gentrification in the area risks 'pricing out' the local community. In 2016, residents' associations protested against plans to redevelop some social housing in the area by private developers, fearing commercialisation would drive out existing residents.

For many students, two case studies should be sufficient to answer this question, but this student makes use of a third (Brick Lane) to highlight further problems that might arise from place-shaping processes – the prospect of higher housing costs for the community at the heart of the place.

To conclude, community groups are frequently the most important players in place-shaping. This is especially true when communities are threatened by direct change or change organically as a result, for example in Coin Street, London. However, in many cases change is led by planners, investors and local or national government, especially in places which have experienced widespread economic change as a result of deindustrialisation, such as Salford Quays, Greater Manchester.

Evaluative language is present throughout this answer, with several links to the quotation in the question. The conclusion is summative, with a clear statement of 'extent to which', i.e. 'community groups are frequently the most important players'. The conclusion also provides a caveat which suggests that 'the extent to which' can be a little more nuanced.

Contemporary urban environments

Question 1

Giving examples, explain the concept of a 'world city'.

(4 marks)

1 mark per valid point.

Student answer

Examples of world cities include London, New York, Milan and Hong Kong ✓. World cities have grown largely due to their global influence, often in terms of financial impact ✓, but also in terms of being centres of education, law, advertising, media, sport and cultural and political strength ✓. They also have high levels of connectedness, both physically (such as having airport hubs) and electronically ✓.

4/4 marks awarded The student provides a number of correct statements.

Question 2

Figure 12 shows aspects of global waste generation, 2010 and 2025 (projected).
Analyse Figure 12.

(6 marks)

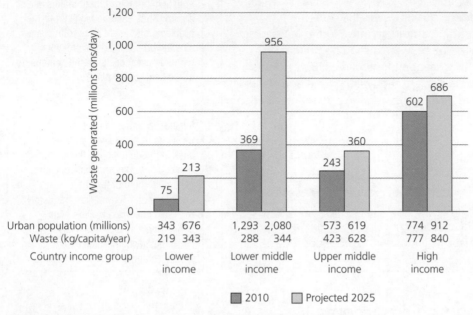

Figure 12 Global waste generation – 2010 and 2025 (projected)

Source: City Lab

Level 2 (4–6 marks)

AO3 – Clear analysis of the quantitative evidence provided, which makes appropriate use of evidence in support. Clear connection(s) between different aspects of the evidence.

Level 1 (1–3 marks)

AO3 – Basic analysis of the quantitative evidence provided, which makes limited use of evidence in support. Basic connection(s) between different aspects of the evidence.

Student answer

Figure 12 shows that global waste generation will increase all around the world, in every type of country. In 2010, the largest amount of waste came from high-income countries (HICs), about 46% of all waste, but by 2025 it is expected that most waste will come from lower middle income countries (LMICs), about 43%. This means that the rate of increase in the generation of waste has more than doubled in LMICs whereas there has only been a small increase in HICs.

However, if we look at the generation of waste on a per capita basis, we can see that individuals in HICs will generate more than twice as

6/6 marks awarded In this answer, the student recognises that although the data may initially appear to be straightforward, it is in fact extremely complicated. He/she starts with a simple statement, before outlining some significant changes that will have taken place over the time period. Note the use of both quantitative and qualitative statements. The student enters Level 2 at this stage of the answer.

much waste (840 kg/capita/year) as individuals in each of lower-income countries (LICs) and LMICs (343/344 kg/capita/year respectively). This is despite the fact that urban populations in LMICs will be twice the size of urban populations in HICs.

Finally, it is worth noting that the rate of increase in the generation of waste is greatest in LICs – the figure for 2025 is almost three times that of 2010. This could be linked to the doubling of the urban population here.

The second and third paragraphs continue this excellent work, each highlighting a number of connections in the data.

It is clear that the issue of urban waste generation is set to cause problems in the near future.

The student ends with a concluding sentence which, although valid, is not needed for this type of question.

Question 3

Assess the extent to which environmental problems in one city you have studied are a consequence of globalisation.

(9 marks)

Level 3 (7–9 marks)

AO1 – Demonstrates detailed knowledge and understanding of concepts, processes, interactions and change. These underpin the response throughout.

AO2 – Applies knowledge and understanding appropriately with detail. Connections and relationships between different aspects of study are fully developed with complete relevance. Evaluation is detailed and well supported with appropriate evidence.

Level 2 (4–6 marks)

AO1 – Demonstrates clear knowledge and understanding of concepts, processes, interactions and change. These are mostly relevant though there may be some minor inaccuracy.

AO2 – Applies clear knowledge and understanding appropriately. Connections and relationships between different aspects of study are evident with some relevance. Evaluation is evident and supported with clear and appropriate evidence.

Level 1 (1–3 marks)

AO1 – Demonstrates basic knowledge and understanding of concepts, processes, interactions and change. This offers limited relevance with inaccuracy.

AO2 – Applies limited knowledge and understanding. Connections and relationships between different aspects of study are basic with limited relevance. Evaluation is basic and supported with limited appropriate evidence.

Student answer

The population of Delhi, India, is over 20 million. It is estimated that each year the city's population increases by over 300,000 as a result of inward migration and by an additional 250,000 as a result of natural increase; this has made Delhi one of the fastest-growing cities in the world. Globalisation has affected its growth greatly, and accounts for many of the city's environmental problems.

Delhi is the economic and administrative hub of the country. It is also an important light industrial centre, with over 130,000 industrial units, producing everything from TVs to medicines. Many multinationals have moved into the city. For example, Delhi has emerged as the fashion capital of India – more than 60% of the design community resides here.

It is estimated that 52% of Delhi's population live in slum conditions. In comparison to Delhi's average infant mortality rate (IMR) of 40, the IMR in slums is higher at 54 for every 1,000 live births. There is a high incidence of diseases, such as diarrhoea, because of the lack of safe drinking water and sanitation. 31% of Delhi's slum dwellers have no sanitation facilities. Large-scale migration driven by globalisation may have caused these problems, though some could argue that dirty water is the fault of poor sanitation resulting from poor governance within the city.

Delhi has severe air pollution. By 2018, the city's average PM_{10} level was very high, more than four times the level the World Health Organization considers safe. One in three Delhi residents has at least one respiratory problem as a result of this. The Yamuna is one of the most polluted rivers in the world, especially around Delhi, which dumps about 58% of its waste into the river. Delhi alone contributes 3.2 million litres per day of sewage to the river. The problem is aggravated by the fact that the water remains stagnant for almost nine months of the year – until the summer monsoon arrives. Air and water pollution are direct consequences of the processes of globalisation such as industrialisation.

It is therefore clear that globalisation has caused significant environmental problems in Delhi, India.

7/9 marks awarded Note that this synoptic question connects the content areas of Contemporary urban environments and Global systems. The student begins with two paragraphs that set the scene for the answer to come. Some good case study detail is provided here.

The third and fourth paragraphs then discuss two environmental problems (sanitation and pollution) within Delhi, again with good detail, and each paragraph ends with an explicit link to the theme of the question, globalisation. Such depth of knowledge and clarity of linkage takes the answer into Level 3.

The answer ends with a brief, but necessary, summative sentence. The answer could have been improved with a few more direct links to the role of globalisation in the city's environmental problems.

Question 4

Evaluate the impact of a river restoration and conservation project in a damaged urban catchment you have studied on the character of that place.

(9 marks)

Level 3 (7–9 marks)

AO1 – Demonstrates detailed knowledge and understanding of concepts, processes, interactions and change. These underpin the response throughout.

AO2 – Applies knowledge and understanding appropriately with detail. Connections and relationships between different aspects of study are fully developed with complete relevance. Evaluation is detailed and well supported with appropriate evidence.

Level 2 (4–6 marks)

AO1 – Demonstrates clear knowledge and understanding of concepts, processes, interactions and change. These are mostly relevant though there may be some minor inaccuracy.

AO2 – Applies clear knowledge and understanding appropriately. Connections and relationships between different aspects of study are evident with some relevance. Evaluation is evident and supported with clear and appropriate evidence.

Level 1 (1–3 marks)

AO1 – Demonstrates basic knowledge and understanding of concepts, processes, interactions and change. This offers limited relevance with inaccuracy.

AO2 – Applies limited knowledge and understanding. Connections and relationships between different aspects of study are basic with limited relevance. Evaluation is basic and supported with limited appropriate evidence.

Student answer

One river restoration and conservation project I have studied in a damaged urban environment is Manor Ponds in Sheffield. It is a sustainable urban drainage scheme (SuDS). This SuDS scheme is located on council-owned public open space (2 hectares), next to a regenerated housing area, Manor Farm.

The housing area exists on a steep slope. The existing main water sewer nearest to the SuDS site was inaccessible as it lay above it, within the housing estate. So, the water had to be drained down the slope. The SuDS site is being developed as an inner city park and therefore releasing water into the park in a high-quality state was seen as a basic requirement, as the character of it being a place for children to play and others to walk the dog safely was important.

One of the main challenges of the project was the initial character of the place – a rundown area with high levels of deprivation. In the early stages of construction, the site experienced problems with crime including vandalism, such as dumping and burning out of cars.

9/9 marks awarded Note that this synoptic question connects the content areas of Contemporary urban environments and Changing places. This is an excellent answer, with clear detail of the chosen area of study and regular links to the thrust of the question – the impact on the character of a place. The opening paragraph sets the scene for the answer that follows.

The natural assets of the park including grassland and wetlands have been retained. The need for drainage in order to provide better access and recreation opportunities has been carried out with the proviso that water should remain visible and not hidden. Open water was therefore seen as an asset in creating character in the site. Grass turf was used to cover vulnerable areas of the system such as overflow channels and some slopes. This all created a pleasant 'green feel' to the character of the place.

In summary, the project has been able to reclaim derelict land into an improved landscape. It constructed a recreational space for the housing area and enhanced the wetland ecosystem of the area. Not only did the park provide a recreational area, it also provided an educational opportunity for all the community. The character of the place was therefore enhanced.

> The second, third and fourth paragraphs each link specific aspects of the chosen location, and scheme, to 'place'.

> The conclusion neatly rounds off the answer, giving both the 'big picture' and a clear statement of evaluation.

Question 5

To what extent do urban areas modify their climate?

(20 marks)

See the generic mark scheme on page 138.

Student answer

Urban areas differ from their surrounding rural areas greatly as their temperature, precipitation amounts, wind speed and air quality are all affected. So, it is true that urban areas do modify their climate. I shall examine each of these aspects in turn.

Temperature in urban areas is generally 1 or 2°C warmer than the surrounding rural areas. This is due to the type of building materials used. The majority of buildings are made of dark, matt materials with a low albedo which absorb heat during the day, and release it at night warming the atmosphere. Night time temperatures can be as much as 5 to 10°C warmer than rural temperatures. Some modern building materials such as glass have a high albedo, and therefore reflect heat downwards on to the streets, also increasing air temperatures. In addition to building materials, industry and even people of which there are high densities in urban areas, emit heat. Air conditioning units, while cooling indoor temperatures, release heat into the atmosphere also.

> **18/20 marks awarded** This is a very methodical answer, which is both conceptually thorough and logical. The student begins with a clear statement of intent.

In addition to temperature, precipitation in urban areas is also increased. Anthropogenic heat, described above, leads to relatively low pressure areas developing over cities. The warm air rises due to its low density and eventually cools and condenses to form clouds and subsequent precipitation. In cities with significant air pollution problems, particulates released from industry and vehicle exhausts act as condensation nuclei for water droplets and therefore increase the rate of cloud formation. Urban areas can create their own thunderstorms due to the intense heat causing air to rise rapidly, forming huge cumulonimbus clouds upon condensation. These thunderstorms generally occur in the late afternoon or evening as the ground has received a large amount of insolation from the sun throughout the day.

Conversely to temperatures and precipitation, the wind speed in urban areas is generally decreased. It can be up to 5% lower in the city centre than the suburbs due to the high density of uneven building heights. This increases the frictional drag between the air flow and tops of buildings, slowing the wind speed. However, wind speed can be much higher in some parts of cities due to the venturi effect in urban 'canyons' which are long narrow streets parallel to the direction of airflow. The venturi effect is the effect of squeezing the airflow, which produces high velocities. These high wind speeds can sometimes blow over pedestrians. Bridgewater Place, the tallest building in Leeds, produces an unusual spiralling effect of the air around it, which blew over a lorry killing a man in 2009.

Air quality is a lot worse in urban areas, as factories and vehicles produce large amounts of pollutants such as carbon dioxide, sulphur dioxide and PM_{10}s. These chemicals create a pollution dome, which increases warming of the urban microclimate, as they allow in solar radiation but trap outgoing thermal infrared radiation, preventing it from being radiated back into space. Photochemical smog formed by ultraviolet light reacting with hydrocarbons and oxides, produced by vehicle exhausts, is common.

The following paragraphs then examine each aspect of climate (temperature, precipitation, wind speed and air quality) stated in the introduction with clear detail of process, and appropriate application to the question. One piece of case study material is also well-used – Bridgewater Place, Leeds.

It is clear therefore that urban areas do significantly modify their climates, in comparison to surrounding rural areas. Most of the modifications are not too damaging to people, with the exception of air pollution. Although there are many policies in place to reduce the effects of pollution in urban areas, these policies cannot reduce these effects to zero because of the high concentration of industry in cities. Global economic systems and the character of the urban places themselves therefore also play an important role.

The lengthy conclusion addresses the question, but also introduces other factors that may play a role – a clear attempt to introduce synoptic ideas – governance, global systems and changing places. A little more consideration of these elements earlier would have raised the answer to full marks.

Population and the environment

Question 1

Give two reasons for the growth in the number of female migrants globally in the twenty-first century.

(4 marks)

1 mark per valid point.

Student answer

There is greater independence and/or freedom of women today in some countries than in the recent past ✓. This is largely a result of less discrimination in the labour market in those countries; for example, women's rights are better respected in some countries such as USA and UK ✓. Another factor is family reunification where women, and their children, join earlier male migrants after they have become established in the host country ✓, perhaps now with accommodation and possible employment in a family-run business✓.

4/4 marks awarded The student provides four correct statements. Note the final point is a development of the preceding point.

Question 2

Study Figure 13 which shows countries by Age Dependency Category (2018). Analyse Figure 13.

(6 marks)

Figure 13 Countries by Age Dependency Category (2018)

Source: Population Reference Bureau (2018)

Level 2 (4–6 marks)

AO3 – Clear analysis of the qualitative evidence provided, which makes appropriate use of evidence in support. Clear connection(s) between different aspects of the evidence.

Level 1 (1–3 marks)

AO3 – Basic analysis of the qualitative evidence provided, which makes limited use of evidence in support. Basic connection(s) between different aspects of the evidence.

Student answer

The countries that have a high old-age dependency are entirely in the northern hemisphere, for example Canada, Russia, the UK, Eastern European countries and China. All of these are developed nations in Stage 4 or 5 of the Demographic Transition Model (DTM), except China which is an emerging economy.

Africa dominates the high child dependency distribution, except its north (Libya) and south (South Africa). This area is mostly in Stage 2 of the DTM. Other such countries are in southwest Asia (Pakistan, Afghanistan) and one in South America (Bolivia).

Interestingly, both the USA and Australia have a double dependency – high child and old-age dependency. This is probably due to immigration. Argentina is also in this category, so its distribution is very spread.

Only one country seems to have a low overall dependency. This country is near Saudi Arabia – I think this is either Oman or Yemen.

6/6 marks awarded Answering such questions may appear to be straightforward (assuming you know the world's countries), but this is not always the case. The key is *not* to provide explanation, tempting as it may be. Here the student does provide material to try and explain patterns of dependency, such as references to the DTM and immigration. However, credit can be awarded here for each of the analyses of high old-age (in the first paragraph), high child (in the second paragraph) and double dependency (in the third paragraph). Note the additional credit for the correctly identified anomalies in Africa (in the second paragraph).

Credit can also be awarded for the final sentence because of the qualitative use of 'Only', and the first named country of the two identified is correct.

Question 3

Evaluate how globalisation has contributed to patterns of population change in a country or society you have studied.

(9 marks)

Level 3 (7–9 marks)

AO1 – Demonstrates detailed knowledge and understanding of concepts, processes, interactions and change. These underpin the response throughout.

AO2 – Applies knowledge and understanding appropriately with detail. Connections and relationships between different aspects of study are fully developed with complete relevance. Evaluation is detailed and well supported with appropriate evidence.

Level 2 (4–6 marks)

AO1 – Demonstrates clear knowledge and understanding of concepts, processes, interactions and change. These are mostly relevant though there may be some minor inaccuracy.

AO2 – Applies clear knowledge and understanding appropriately. Connections and relationships between different aspects of study are evident with some relevance. Evaluation is evident and supported with clear and appropriate evidence.

Level 1 (1–3 marks)

AO1 – Demonstrates basic knowledge and understanding of concepts, processes, interactions and change. This offers limited relevance with inaccuracy.

AO2 – Applies limited knowledge and understanding. Connections and relationships between different aspects of study are basic with limited relevance. Evaluation is basic and supported with limited appropriate evidence.

Student answer

One of the main outcomes of globalisation is international migration, which greatly impacts the pattern of population change in a country. One example is the movement of Polish people to the UK in the twenty-first century to find employment. Poland has dominated the movement of people from the EU into the UK since 2004 – in 2015 there were over 800,000 people born in Poland resident here.

9/9 marks awarded Note that this synoptic question connects the content areas of Population and the environment and Global systems. The student begins by clearly identifying the country affected together with a clear link to the thrust of the question.

The reasons for the flow of Polish migrants to the UK are varied, with social, political and economic factors playing a part. These are all features of increasing globalisation. Unemployment in Poland in 2004 was close to 20%, compared to 5% in the same year in the UK. The majority of Polish migrants are of working age, and are often relatively highly educated. Many of these migrants are single workers who send remittances home and/or save money to buy a property or set up a business when they return to Poland.

There is a significant financial benefit for Polish nationals to migrate to the UK in search of work. Wage rates here are higher than those back in Poland. The higher welfare benefits for families in the UK compared to Poland also act as a significant pull factor. Other pull factors are social and cultural – many Poles have migrated to the UK to be with family members or to study. Settling-in is made easier by the well-established Polish communities in parts of London, Doncaster and Peterborough. Newcomers can tap into existing support networks, some dating back to the 1940s when Polish aircraft pilots were stationed in the UK during WW2 (an early form of globalisation?) This link is continued today with the use of the East European airline WizzAir to get here.

The following two paragraphs demonstrate detailed knowledge of the case study, together with explicit links to the question.

The movement of Polish people into the UK has clearly been a consequence of globalisation. However, after the Brexit vote in 2016, there are signs that many such people are returning to their home country. The reasons for this are more social rather than economic, and perhaps a reflection of the growing unpopularity of globalisation in the world today.

The conclusion provides a summative statement of evaluation, as required by the question, together with an up-to-date qualification. Such use of contemporary information will always impress an examiner.

Question 4

'Poverty is the main factor in the spread of infectious disease.' How far do you agree with this statement?

(9 marks)

Level 3 (7–9 marks)

AO1 – Demonstrates detailed knowledge and understanding of concepts, processes, interactions and change. These underpin the response throughout.

AO2 – Applies knowledge and understanding appropriately with detail. Connections and relationships between different aspects of study are fully developed with complete relevance. Evaluation is detailed and well supported with appropriate evidence.

Level 2 (4–6 marks)

AO1 – Demonstrates clear knowledge and understanding of concepts, processes, interactions and change. These are mostly relevant though there may be some minor inaccuracy.

AO2 – Applies clear knowledge and understanding appropriately. Connections and relationships between different aspects of study are evident with some relevance. Evaluation is evident and supported with clear and appropriate evidence.

Level 1 (1–3 marks)

AO1 – Demonstrates basic knowledge and understanding of concepts, processes, interactions and change. This offers limited relevance with inaccuracy.

AO2 – Applies limited knowledge and understanding. Connections and relationships between different aspects of study are basic with limited relevance. Evaluation is basic and supported with limited appropriate evidence.

Student answer

A wide range of factors affect patterns of infectious diseases – of these, poverty is highly significant. The relatively poor governments of low-income developing countries (LIDCs) and some emerging developing countries (EDCs) may be unable to provide sufficient healthcare resources for their populations. Numbers of hospitals, local medical facilities, trained health workers and medicines such as vaccines may be in short supply simply because of costs. As a result, populations most affected by the spread of viruses are rural communities which have limited accessibility, for example, the rural and urban poor in sub-Saharan Africa and parts of South East Asia.

Viruses spread rapidly where people live in close contact in overcrowded conditions or where there is poor sanitation. Inadequate diet leading to under-nutrition and malnutrition in poor families weakens immunity and also contributes to rapid spread of disease following an outbreak. For example, influenza, diarrhoeal disease, malaria and TB are the major causes of death from infectious disease in Malawi.

8/9 marks awarded Note that this synoptic question connects the content areas of Population and the environment and Global systems. The answer begins with a comprehensive overview of the role of poverty in the spread of infectious diseases.

This is continued in the second paragraph, where a little more specificity is provided.

On the other hand, a range of factors other than poverty that influence patterns of infectious disease can be illustrated by the Ebola outbreak in West Africa in 2013. The initial inability of medical staff to recognise the disease was significant – some health workers and doctors became infected and their subsequent attempts to assist in outlying settlements spread the virus further. Daily population movements to local markets in the neighbouring countries of Guinea, Liberia and Sierra Leone also spread the virus rapidly. Traditional cultural customs of washing, kissing and touching at funerals caused further transmission of the disease. Years of civil war in Sierra Leone destroyed infrastructure, including medical resources. Also many local communities were suspicious of outside intervention, which added to the delay in specialist help from advanced countries (ACs).

It is fair to state that poverty does have a significant impact on disease patterns, especially in LIDCs, which are at an early stage of the epidemiological transition. However, this is just one of many social, economic, political and environmental factors that contribute to the spread of infectious disease.

The third paragraph provides an alternative perspective, with much more detailed use of case study material.

The final paragraph provides an overall statement of assessment, as required, together with a general recognition that factors other than poverty may play a role. The answer could have been improved with a little more specific detail being provided in the second and third paragraphs.

Question 5

Evaluate the effectiveness and sustainability of techniques used to improve food security.

(20 marks)

See the generic mark scheme on page 138.

Student answer

I believe that modern-day techniques and methods to improve food security, such as genetic modification (GM) and food banks, can be both effective and sustainable, and that without such new ideas, global population growth would exceed food supply.

Firstly, the rise of GM crops has created a better likelihood of plant survival and being drought resistant. In China, land is being degraded rapidly and the introduction of GM crops has meant that more plants can be grown per square metre and they are better adapted to difficult and changeable environments. Despite being expensive, GM creates a reliable and sustainable food source that can feed more of China's growing population. Controversy has arisen over the ethical issues of GM crops, yet despite this, many thousands of people do not care about genetic manipulation but they do need a reliable food source to feed their families and generate profits at markets.

20/20 marks awarded The first thing to note about this answer is that the student has written in a very personalised way, making use of the first-person. Such an approach is perfectly acceptable even though some of the opinions given may seem a little direct. The introduction is brief, but clearly linked to the question.

Non-governmental organisations (NGOs) have been crucial in ensuring food security. NGOs, such as the Trussell Trust, have set up food banks to ensure poor people in the UK do not starve. This is effective, yet not sustainable, as the 3.4 million living in poverty in the UK could become dependent on the aid. During the 2010–15 Coalition government in the UK, a policy was introduced to give every primary school child in the first three years of school a free school dinner. The success of this wasn't fully realised until 2016, during the summer holidays, when hundreds of children were going hungry in Wales and some school kitchens had to reopen. This shows that children must have a nutritious meal a day which aids learning and will eventually lift them out of poverty as they get better jobs.

Elsewhere, NGOs in the Sahel are providing farmers with acacia seeds to start reducing desertification through nitrogen-fixing trees as well as allowing them to sell gum, and peanuts, on international markets. This creates a sustainable income for those in the Sahel and the improvement of desertification means that eventually more crops will be able to be grown, especially with NGO support and training in fallow periods and livestock management.

New innovative ways to farm have sprung up like the organoponics project in Cuba, which has helped to create food security in Havana. This allows people to grow food in small concrete blocked areas, filled with organic-rich soil, in the city. This is sustainable, as many in Cuba live in urban slum settlements and they need fresh vegetables to improve nutrition levels.

In Kibera, Kenya, a similar gardening project gives food security to those in the slums, as the contaminated soils are toxic and they cannot grow the food they need. It also puts profits into a trust for its workers to start their own businesses, further alleviating the food insecurity problem. This, I believe, is one of the most effective ways of helping the world's hungry to be self-dependent and not rely on aid for their food.

The following five paragraphs each provide clear and/or detailed accounts of a wide range of strategies to improve food security around the world. The breadth of knowledge is impressive, and note that in each paragraph, an explicit link to the question is provided.

I believe that today we are providing those in developing countries with the chance to become food secure themselves in a way that is both sustainable and effective for years to come. NGOs and GM crops, as well as methods like urban farming, are alleviating poverty and reducing dependence on international aid. To me, despite having a long way to go, the recent advancements in yields and ingenious ways of growing plants have made for an effective and sustainable culture surrounding the future of food in the world.

The conclusion is also very effective, with a strong sense of evaluation. This is an excellent answer.

Resource security

Question 1

Giving examples, explain the difference between primary and secondary sources of energy.

(4 marks)

1 mark per valid point.

Student answer

Primary energy sources are those used for power in their natural or unprocessed form ✓. They include wood, fossil fuels (oil, natural gas and coal) ✓ and power from uranium (nuclear), as well as renewable sources such as solar or tidal ✓.

Some of these sources may be manufactured or converted into secondary sources – such as electricity, from burning coal and gas, for commercial, industrial and domestic use ✓.

4/4 marks awarded The student provides four correct statements.

Question 2

Table 5 provides information on the world's top ten coal-producing countries, and coal consumption in the top ten coal-consuming countries in 2015.

Table 6 shows a Spearman's rank correlation calculation used to study whether there is a relationship between production and consumption of coal in the countries shown in Table 5.

Analyse the data shown in Tables 5 and 6.

(6 marks)

Table 5 Information on coal-producing and coal-consuming countries in 2015

Coal-producing country by rank	Coal production (million tonnes oil equivalent)	Coal-consuming country by rank	Coal consumption (million tonnes oil equivalent)
1 China	2,480	1 China	1,920
2 USA	764	2 India	407
3 India	361	3 USA	396
4 Australia	278	4 Japan	119
5 Indonesia	247	5 Russia	89
6 Russia	229	6 South Africa	85
7 South Africa	176	7 South Korea	84
8 Germany	133	8 Indonesia	80
9 Poland	91	9 Germany	78
10 Kazakhstan	87	10 Poland	50

The null hypothesis is, 'There is no relationship between the production of coal and the consumption of coal.'

Table 6 Spearman rank calculation for the countries shown in Table 5

Rs value	0.806
Critical value at the 0.05 significance level	0.648

Level 2 (4–6 marks)

AO3 – Clear analysis of the quantitative evidence provided, which makes appropriate use of evidence in support. Clear connection(s) between different aspects of the evidence.

Level 1 (1–3 marks)

AO3 – Basic analysis of the quantitative evidence provided, which makes limited use of evidence in support. Basic connection(s) between different aspects of the evidence.

Student answer

From Table 6 it can be concluded that there is, with 95% certainty, a correlation between production and consumption of coal as the calculated Spearman's rank value is above the critical value given. This correlation is positive as indicated by the positive SR value and is arguably strong since the value of 0.806 is near to 1.

China greatly dominates both coal production and consumption – in 2015, it produced 2,480 million tonnes oil equivalent, over 45% of the total produced by the top ten producers that year. It also consumed 50% of total consumption by the top ten consuming countries. The correlation is not perfect, as India consumes a higher amount of coal than the USA, but produced less than half of the USA's total. There are few countries on either list that do not feature on the other, for example South Korea in consumption but not production.

Overall, there is certainly a strong correlation between the two, and it is dominated by China.

6/6 marks awarded The student addresses the correlation in Table 6 first, recognising the significance and strength of the correlation. The answer then moves on to the data in Table 5 and makes a number of good qualitative and quantitative comments. The final point, re South Korea, is a little simplistic, but by this point top Level 2 credit can be awarded.

Question 3

Study Table 7. Using Table 7, and your own knowledge, assess the costs and benefits of changes to a country's water supply and security.

(9 marks)

Table 7 Information about Singapore's population, water use and sources of water supply (2016 and 2060)

Year	Total population (millions)	Water use per capita per day (litres)	NEWater: (Recycled, treated and grey water)	Desalination	Trans-boundary water imports from Malaysia	Local rainfall catchments
2016	5.69	150	30%	10%	40%	20%
2060 (planned)	6.56	130	55%	30%	–	15%

Level 3 (7–9 marks)

AO1 – Demonstrates detailed knowledge and understanding of concepts, processes, interactions and change. These underpin the response throughout.

AO2 – Applies knowledge and understanding appropriately with detail. Connections and relationships between different aspects of study are fully developed with complete relevance. Evaluation is detailed and well supported with appropriate evidence.

Level 2 (4–6 marks)

AO1 – Demonstrates clear knowledge and understanding of concepts, processes, interactions and change. These are mostly relevant though there may be some minor inaccuracy.

AO2 – Applies clear knowledge and understanding appropriately. Connections and relationships between different aspects of study are evident with some relevance. Evaluation is evident and supported with clear and appropriate evidence.

Level 1 (1–3 marks)

AO1 – Demonstrates basic knowledge and understanding of concepts, processes, interactions and change. This offers limited relevance with inaccuracy.

AO2 – Applies limited knowledge and understanding. Connections and relationships between different aspects of study are basic with limited relevance. Evaluation is basic and supported with limited appropriate evidence.

Student answer

Table 7 shows that Singapore plans to shift its water supply from imports from Malaysia, towards water recycling and desalination by 2060. At the same time, its population will rise by close to 1 million, but water use per person falls by 20 litres per day.

There is a potentially very large benefit of ending use of imported water from Malaysia. In 2016, 40% of Singapore's water came from another country that could reduce or stop the supply. There are examples where water-sharing agreements lead to tensions, such as in the Nile Basin, so relying on another country for water is potentially risky for water security. However, if the relationship is just an economic one, i.e. Singapore pays Malaysia for water, it could be sustainable.

Recycled grey water, as a form of water conservation, has the benefits of using the same water multiple times, so reducing demand for new

9/9 marks awarded This is an excellent answer to this data stimulus question. It refers to all of the data in Table 7, reference to which is essential. The introductory paragraph makes direct reference to the data and shows good understanding of it. At the start, there is a clear judgement about the significance of the benefit of Singapore reducing its reliance on water from Malaysia. Some brief reference is made to other trans-boundary water situations which shows breadth of understanding. Assessment is shown by the counterarguments that getting water from Malaysia may not actually be very insecure at all.

supply. Water conservation is also evident in the aim to reduce per capita daily consumption. In the UK this is achieved by more efficient washing machines, dishwashers and showers. It may have the benefit of reducing water bills, as well as preventing increases in water demand.

The shift to 30% of water from desalination has the benefit that Singapore will control this water supply, rather than Malaysia controlling it. However, desalination has costs. It is expensive to build the plants, and these need a large energy source such as oil or natural gas to run. This means desalination is usually not eco-friendly and has high greenhouse gas emissions. However, it does ensure water security.

The use of rainfall catchments is set to decline by 5%. This might actually mean the volume of water they supply is similar to 2016, because despite the 150 to 130 litre reduction in per person water use, total water demand will rise because of the 800,000 increase in population.

Overall, by 2060 Singapore will have the significant benefit of being much more water secure. However, this could come at the economic costs of higher water bills and some environmental costs from desalination.

> The paragraph on grey water recycling/ NEWater also covers costs and benefits, showing that the issue is being considered from both sides. Environmental and economic costs are considered in relation to desalination, as well as benefits in terms of a secure water supply.

> The section on rainfall catchments is very analytical, i.e. it unpicks that data to understand the changing picture of Singapore's water demand and shows good understanding.

> When the command is 'assess', a final conclusion is always useful, in this case stressing the primary benefit of greater water security for Singapore despite some potential costs.

Question 4

Evaluate the extent to which the increasing demand for energy is the most important factor modifying the global carbon cycle.

(9 marks)

Level 3 (7–9 marks)

AO1 – Demonstrates detailed knowledge and understanding of concepts, processes, interactions and change. These underpin the response throughout.

AO2 – Applies knowledge and understanding appropriately with detail. Connections and relationships between different aspects of study are fully developed with complete relevance. Evaluation is detailed and well supported with appropriate evidence.

Level 2 (4–6 marks)

AO1 – Demonstrates clear knowledge and understanding of concepts, processes, interactions and change. These are mostly relevant though there may be some minor inaccuracy.

AO2 – Applies clear knowledge and understanding appropriately. Connections and relationships between different aspects of study are evident with some relevance. Evaluation is evident and supported with clear and appropriate evidence.

Level 1 (1–3 marks)

AO1 – Demonstrates basic knowledge and understanding of concepts, processes, interactions and change. This offers limited relevance with inaccuracy.

AO2 – Applies limited knowledge and understanding. Connections and relationships between different aspects of study are basic with limited relevance. Evaluation is basic and supported with limited appropriate evidence.

Student answer

Global energy demand is expected to increase by 40% to 2030. Fossil fuels supply more than 80% of the world's energy and this is unlikely to change soon, despite increases in renewable energy. About 70% of carbon dioxide production worldwide is from burning fossil fuels. This has increased the concentration of carbon dioxide in the atmospheric store from 315 ppm in 1958 to over 410 ppm today.

Not all energy sources are the same. Some sources – such as nuclear and wind power – are low-carbon, but these do not meet most of global energy demand. Coal and oil are the main sources of energy. In the near future, the development of Canada's tar sands and the USA's oil shales are likely to release even greater amounts of geological carbon. This is because a lot of energy is used in extracting and processing these unconventional fossil fuels even before they are used – so emissions are far greater. Growing global energy demand makes it likely that these sources will be tapped in order to increase supply.

Land-use change, especially deforestation, is responsible for up to 25% of global greenhouse gas emissions. Up to 8 million hectares of tropical forest have been destroyed in Indonesia alone just for palm oil. However, deforestation is at least reversible because forests can be replanted.

Farming is a major contributor to atmospheric greenhouse emissions, with 80% of all methane emissions from farming, including farm animals and wet rice cultivation. As the global population climbs toward 9 billion by 2050, the challenge of feeding these additional people is likely to increase emissions further, as well as leading to even more deforestation to convert forests to food production.

Overall, energy demand does have the greatest impact on modifying the carbon cycle, by releasing fossil carbon into the atmosphere and oceans. Deforestation and farming are two other factors, which are often linked. While less significant than energy demand, their importance is only likely to grow as there are more mouths to feed, whereas it may be possible to reduce fossil fuel use with renewable energy technologies in the future.

9/9 marks awarded This is an excellent answer to this synoptic question, which links Resource security with the Carbon cycle. The introduction shows good understanding of energy demand and how this leads to emissions which alter the carbon cycle.

The following paragraph judges that some energy sources are more significant than others –both today and in the near future.

The student also considers other factors, particularly deforestation. The answer judges this as less significant than fossil fuel burning because deforestation can be stopped and even reversed.

Farming is also considered and he/she argues that this threat is likely to grow in the future.

Throughout, examples and data are used to support the answer, i.e. evidence is presented. There is also a coherent overall conclusion.

Question 5

With reference to one case study, evaluate the environmental impacts of a major water supply scheme incorporating a major dam and/or barrage and associated distribution networks.

(20 marks)

See the generic mark scheme on page 138.

Student answer

China intends to complete a $60 billion South–North water transfer project by 2050. The project will divert 45 billion m^3 of water per year from the Yangtze and Han Rivers in southern China to the Yellow River Basin in arid northern China. The scheme will displace hundreds of thousands of people, and it will also have a significant environmental impact across the country.

Northern China has long been a centre of population, industry and agriculture and with all three growing rapidly, the per capita share of the region's water resources has inevitably kept falling. Historically, this has led to the over-exploitation of groundwater – often supplying urban and industrial development at the expense of agriculture, leading to severe water shortages in rural areas. In addition, land subsidence and the region's frequent sandstorms have also been linked to the excessive use of groundwater.

Many are concerned that the project could exacerbate water pollution problems. Pollution from factories may render the water unfit to drink. Meanwhile, the diversion of water from the Yangtze River Basin to the north is likely to exacerbate pollution problems on the Yangtze – problems that have worsened since the construction of the Three Gorges Dam.

As with China's other mega-project, the Three Gorges Dam, the new diversion scheme has provoked many environmental concerns, mainly the displacement of people and the destruction of ecosystems. There are worries that the project will change further the ecosystems of the Han and Yangtze Rivers. There is evidence that it has already interrupted fish migration and altered the river's chemical balance, temperature and flow velocity.

Environmentalists fear that populations of the Chinese sturgeon and Chinese paddlefish have been affected and may become endangered. Commercial fisheries in the Han River and off the river's mouth in the East China Sea may decline sharply.

20/20 marks awarded This is a clear and detailed answer. From the outset, the student clearly identifies the chosen case study, and uses the introduction to set the scene for what is to follow.

The second paragraph also provides some background to the case study, cleverly referring to existing environmental problems in the receiving area.

There are fears that the enlarged Danjiangkou reservoir dam (D-Dam) on the Han River could have the same impacts as the Three Gorges Dam. One of the biggest fears is that the reservoir will turn into one of the biggest sewers in the world. Sewage treatment works have not yet been built and raw sewage from the rapidly developing upstream cities is allowed to flow downstream and become trapped behind the dam where it damages the local biodiversity and causes water pollution.

Sediment transport will be affected too. Over the period of the Three Gorges Dam's construction, river sediment fell by over 50% and by 2004 it was 35% less than average. Now that the dam is complete, even with its sluice gates, it is clear that most of the sediment is still being deposited in the reservoir, potentially blocking the port of Chongqing, decreasing the fertility of downstream soils and leading to the erosion of coastal wetlands. Once again, the same could happen at the new D-dam further north.

In conclusion, it can be seen that the proposed South–North water transfer project could have severe environmental costs. These include water pollution, species loss, interruptions to natural movements of sediment and probable secondary impacts caused by enforced movement of people away from their land. The potential damage is huge, but it is fair to say that if anyone can do the scheme effectively, then the Chinese are perhaps the most capable.

The next five paragraphs then focus on a range of environmental issues, also emphasising their probable impact by comparing the chosen case study with another major water scheme in the past, the Three Gorges Dam. Issues relating to water pollution, species loss, sewage treatment and sediment are examined in turn. The level of detail provided is excellent.

Finally, the conclusion summarises the points previously made, and ends with a degree of optimism. There is one further point that can be made: the student refers to the environmental impact of population displacement on three occasions, and yet does not expand on this point. Although these references do not affect the overall standard of the essay, it is perhaps better not to mention something that isn't developed at some point.

Knowledge check answers

1 The 'shrinking world' can be illustrated by:
   - Up to the mid-nineteenth century, most travel was on horseback – an average speed of 10 mph.
   - Steam trains then moved at up to 70 mph, and steamships at over 30 mph.
   - By the mid-twentieth century, air travel was much more common, again reducing travel times across the world.
   - From the late twentieth century, communication has become almost instant with the internet.

2 The MINT countries are Mexico, Indonesia, Nigeria and Turkey.

3 At the highest level, the top of the service hierarchy is to be found in world cities such as London, New York and Tokyo, which are the major nuclei of global industrial and financial command functions. Other cities of prominence include Frankfurt, Chicago, Paris, Milan and Los Angeles.

4 Just-in-time production is a system designed to minimise the costs of holding stocks of raw materials and components by carefully planned scheduling and flow of these materials and components through the production process. It requires a very efficient ordering system and reliability of delivery. It has given rise to a new term in the transportation of goods – logistics.

5 India has become one of the most attractive locations for the outsourcing of services, in particular the state of Karnataka which houses Bangalore – often known as India's 'Silicon City'. Much of this outsourcing has involved call centre work and software development. Labour costs are much lower, but the workforce is both highly educated and has good use of the English language. India is the second-largest English-speaking human resource in the world and has the world's third-largest 'brain bank', with around 2.5 million technical professionals. There is also a burgeoning middle class of some 250–350 million people with increasing purchasing power.

6 For:
   - A huge potential market of around 500 million people.
   - The combined strength of the members form a powerful trade bloc.
   - Freedom of movement for workers within a wide employment market.

   Against:
   - Poor distribution of EU income particularly as the Common Agricultural Policy (CAP) takes so much of the budget.
   - Over-bureaucracy within the European Commission has brought into question its efficiency.
   - The adoption of some European law has been inconsistent across the Union.

7 For example, in India, Ujjivan Financial Services was established in 2005. The company now has 2.3 million customers, operates in 22 states and is one of the most dominant organisations in Indian microfinancing. Its high loan repayment rate of 99.9% is critical to its success. The scheme uses group lending so that it becomes the collective responsibility of the individuals. As they all may want access to future loans, there is increased awareness that no one individual should default on the loan.

8 An imaginary example of a hub company would be a US-owned TNC which outsources some manufacturing to a South Korean TNC, which in turn has a branch factory in China. The US company also has a branch factory in Mexico, and a subsidiary in Germany. In addition, much of the administrative support for all of these is undertaken in back-office work in India. There is therefore a complex web of interconnections.

9 To maintain international peace and security, and to that end: to take effective collective measures for the prevention and removal of threats to the peace, and for the suppression of acts of aggression or other breaches of the peace.

   To develop friendly relations among nations based on respect for the principle of equal rights and self-determination of peoples.

   To achieve international cooperation in solving international problems of an economic, social, cultural or humanitarian character, and in promoting and encouraging respect for human rights and for fundamental freedoms for all without distinction as to race, sex, language, or religion.

10 The UN plays an integral part in social and economic development through its UN Development Programme (UNDP). UNDP administers the UN Capital Development Fund that helps developing countries grow their economies by supplementing existing sources of capital assistance by means of grants and loans. In addition, the World Health Organization (WHO), UNAIDS, The Global Fund to Fight AIDS, Tuberculosis and Malaria, the UN Population Fund, and the World Bank also play essential roles. The UN annually publishes the Human Development Index (HDI) to rank countries in terms of poverty, literacy, education and life expectancy.

11 NATO is a military alliance established in 1949. It was set up to organise collective defence when members from the North Atlantic area were attacked by an external party. At the time of establishment, the main threat was from the Soviet Union (now Russia), but it has also launched military activity in the Balkans, Afghanistan and Libya. There are currently 28 members.

12 Fairtrade is a social movement whose stated goal is to help producers in developing countries achieve better trading conditions and promote sustainability. Members of the movement advocate the payment of higher prices to exporters, as well as higher social and environmental standards. The movement focuses in particular on commodities, or products, which are typically exported from developing countries to

developed countries, but also consumed in large domestic markets (such as Brazil and India), most notably coffee, cocoa, sugar, tea, bananas, cotton and chocolate.

13 The Antarctic Convergence (AC) marks the location where surface waters of the Southern Ocean moving northward sink below sub-Antarctic waters. The AC is a region of faster water current speeds and strong horizontal gradients in density, temperature and salinity. The AC also marks the location of one of several strong atmospheric jets within the Antarctic Circumpolar Current (ACC), which flows eastward around Antarctica. The AC marks an important climatic boundary in terms of both air–sea fluxes and the heat and salt budgets of the oceans.

14 Krill are small crustaceans and are found in all the world's oceans. Krill are considered an important trophic level connection – near the bottom of the food chain – because they feed on phytoplankton and zooplankton, converting these into a form suitable for many larger animals for whom krill makes up the largest part of their diet. In the Southern Ocean, one species of krill makes up an estimated biomass of over 350 million tonnes. Of this, over half is eaten by whales, seals, penguins, squid and fish each year.

15 The International Whaling Commission (IWC) is the global body charged with the conservation of whales and the management of whaling. The IWC currently has 88 member governments from countries all over the world. All members are signatories to the International Convention for the Regulation of Whaling. The Commission coordinates and funds conservation work on many species of cetacean. The Commission has also adopted a strategic plan for whale-watching to facilitate the further development of this activity in a way that is responsible and consistent with international best practice.

16 Locale refers to a place as a setting for particular practices that mark it out from other places. As well as being a location, place has a physical landscape (buildings, parks, infrastructures of transport and communication, signs, memorials, etc.) and a social landscape (everyday practices of work, education and leisure among others).

17 The phrase 'field of care' refers to what Tuan called the 'affective bond between people and place or setting'. Tuan argued that through experience of a place, the daily activity of living in and moving through specific environments, we come to form attachments to places – they become private places. They are the places where people create interpersonal ties and develop social capital, both of which require extended time spent there as well as material settings.

18 Some writers contend that the process of globalisation highlights, rather than eliminates, place, arguing that distinct differences in place are seen nowhere more clearly than through uneven economic and social development. Using any urban High Street as a context for example, place is not constituted by its own locality, but by its global connections. On a High Street you will see not only shops and offices that connect with the wider world through their ownership and the goods they sell, but also a range of ethnic groups within the people who walk on the pavements. Rather than the idea of a 'local community', a community is built through layered local–global interactions.

19 Location – a point in space with specific links to other points in space. A sense of place – the subjective feelings associated with living in a place.

20 The migrants themselves – fleeing war and persecution; perhaps there were economic migrants there too, seeking a better life

The aid workers handing out food, water and other essential items

Police and troops – seeking to maintain some sort of order

Media reporters and cameramen – informing the world of what was happening

National politicians – trying to find a way to deal with the crisis – there were differing views among them

People who live miles away – wanting to help and pressurising governments

People who live miles away – wanting to 'send them back' and pressurising governments

21 Rebranding – the process of regenerating a city's economy and physical fabric as well as projecting a new, positive urban image to the wider world.

Re-imaging – the process of creating a new perspective on a place as seen by others, at home and abroad.

Regeneration – a combination of physical, economic and social renewal of a city or part of a city.

22 For example, in Africa, several countries are more than 60% urban, with three (Gabon, Libya and Western Sahara) over 80% urban. Some countries, such as Chad and Ethiopia, are less than 25% urban.

23 Economic: loss of jobs; closure of businesses; increase in demand for state benefits; de-multiplier effect. Social: increase in unemployment; out-migration; higher crime levels along with family breakdown and alcohol/drug abuse. Environmental: derelict land; long-term contamination of land areas; deteriorating road infrastructure.

24 The proposed High-Speed Rail network between London, Birmingham, Liverpool, Manchester and Leeds is a planned attempt to connect the poorer city regions to the economic core in London and the South East. Since the 1960s, there has also been considerable investment in regional airports like Newcastle, Manchester and Glasgow.

25 Well-known mega-cities include Tokyo, Jakarta, Istanbul, São Paulo and Cairo. There are other lesser-known mega-cities – for example, Chengdu in China, Ghaziabad, Surat and Faridabad in India, Toluca in Mexico, Palembang in Indonesia and Chittagong in Bangladesh.

26 You can consider:
- its 'assets' – attractions, climate, infrastructure (particularly transport), safety and economic prosperity
- its 'buzz' – combination of the analysis of social media platforms (such as Facebook and Twitter) and print media
- its 'connectedness' – the transport networks to tie it into the world economy such as a major international airport(s), and ideally its own docks. Also home-grown media and communications industries

27 Detroit (Michigan, USA) (2016): Average household income in Detroit was about $25,000, half the national average. Two-thirds of Detroit's residents could not afford basic needs like food and fuel and the poverty rate was 38%. Life expectancy in parts of Detroit is just 69 years, and less than 30% of students graduate from high school. Detroit had the second highest murder rate of any US city. Average house prices in Detroit are about $40,000 and it was estimated that there were 30,000 abandoned homes and 70,000 other abandoned buildings.

28 In London, examples of post-modernist buildings include the Gherkin, 20 Fenchurch Street (the Walkie Talkie building), The Shard and City Hall. The Guggenheim museum in Bilbao, and the Museum of Contemporary Art in Barcelona are Spanish examples.

29 An atmospheric condition in which temperature increases with height rather than the more usual decrease.

30 London has the Congestion Charge – taxing vehicles that travel through central areas – and the Greater London Low Emission Zone (LEZ), within which the most polluting vehicles are required to pay a daily charge for being within that area.

31 An urban storm hydrograph has: a higher peak discharge; steeper rising and recessional limbs; shorter lag time.

32 Most sources of electronic waste are in developed/industrial economies. The main destinations that are known of are countries that are developing. Other destinations that are suspected are mainly in least developed and less industrialised countries. The USA exports its electronic waste all over the world; most of the EU's waste goes to Africa and Asia. Australia, Japan and South Korea export almost all their waste to China and India.

33 The London Olympic site: a 350 ha area of East London was 'cleaned up' and has now become the largest new urban park in Europe, with 100 ha of open land and 45 ha of new habitat. Over 2.2 million m² of soil was excavated, of which nearly half was treated by soil washing, chemical stabilisation, bio-remediation or sorting. 80% of the excavated material was reused on site. A total of 235,000 m³ of contaminated groundwater was successfully treated.

34 The Brundtland Report (1987) was the first to coin the term 'sustainable development'.

35 The best-known sustainable cities include Curitiba (Brazil), Chattanooga (USA), Freiburg (Germany), Copenhagen (Denmark) and Putrajaya (Malaysia). The reasons for increased sustainability vary according to the city, for example, Curitiba for its mass transport and waste recycling programmes.

36 Many people state that the world cannot sustain a population of 7 billion, let alone 9 billion. Others argue that it is more a case of where they are located, and how much they consume – both food and other resources. The average person in the USA consumes almost 50 times more energy than a person in Ghana. The vast majority of greenhouse gases come from the resource-using people of the developed world. The argument is therefore whether to stop the over-consumption of resources by a minority, or reduce fertility in large parts of the developing world, especially in southern Africa, where there is under-consumption of resources. Therefore, the key factors are resource availability and resource consumption.

37 The demographic dividend is a temporary benefit. Once several decades have passed, a large number of older workers stop working (retire). With fewer young workers, due to a low birth rate, to fund care for the elderly, the dependency ratio begins to rise again, this time involving retirement costs, rather than childcare expense – creating a 'demographic debt'. Japan, which moved through the demographic transition ahead of other Asian nations and benefited from an economic boom in the 1960s, funds high-quality care for its current large elderly population. Some say the elderly consume more diapers (nappies) than its babies.

38 Another explanation that has been put forward for the lack of trees on the Canadian Prairie is natural fires caused by lightning, which can spread easily across the flat plains. Equally, humans could have originated some fires in prehistoric times in order to clear land for farming and improve grazing, the outcome being a generally treeless landscape.

39 Some examples of climate-smart agriculture are:
- superior seed varieties that can cope better with climate change
- heat-stress tolerant varieties of the main crop, such as rice and wheat
- levelling of land by lasers to improve water efficiency
- using underground pipes for irrigation, so that less water is lost through evaporation
- use of mobile telephony for accessing weather reports and advice services

40 Where irrigation water is unable to drain away due to poor drainage systems, waterlogging has taken place in the soil. The water table rises up through the soil, bringing salts with it. It is a form of human-induced salinisation.

41 **1** Drip irrigation: in Zimbabwe, US aid agencies have distributed drip irrigation kits to 24,000 small farmers. Designed to save labour and water, and enhance nutrition and improve food security, the average drip irrigation kit is easy to install. It uses far less water than traditional bucket-watering.

2 Stone lines: in parts of Burkina Faso, lines of stones are placed across hillsides. When the rains arrive, the water washes down the slope and is caught by the stones. Any soil carried by the water is also caught.

42 According to the WHO, there are health opportunities that depend on implementing good water practices, which include the following:
- By improving basic water supply, sanitation and hygiene systems, 4% of the global disease burden could be prevented.
- Procedures such as water safety plans (including education and the use of disinfectant) would improve and protect drinking water quality at the community level.
- Increasing availability of simple and inexpensive approaches to treat and safely store water at the household level would improve water supply.

43 Médecins Sans Frontières (MSF) is a worldwide movement, with 90% of its income coming from individual donations, which the organisation says allows it to stay independent and impartial. MSF works in over 60 countries around the world, with specialist teams ready for any health emergency. MSF teams monitor epidemics on the ground continuously and are able to mount rapid emergency responses. For example, in 2014, MSF treated over 47,000 people in 16 cholera outbreaks around the world. They provided beds, plastic sheeting, oral rehydration salts and surgical equipment such as gloves and gowns. They also managed a meningitis outbreak in Niger, where as many as 350 patients were being treated on a daily basis.

44 In-migration can cause a regression within the Demographic Transition Model (DTM) such that a country that was previously in stage 4 or stage 5 could revert to stage 3. Migration tends to consist of younger adults who have families in their 'new' country and hence the birth rate increases again. This has been shown to be the case within the UK in the last 20 years, largely due to the in-migration of young adults from Eastern Europe.

45 Soaring growth rates for emerging economies, e.g. the BRIC (Brazil, Russia, India and China) and MINT (Mexico, Indonesia, Nigeria and Turkey) countries, are creating hundreds of millions of new 'middle-class' consumers. In India alone, this group could grow to become over 250 million by the early 2020s. Will India be able to support the high consumption levels of this middle class? Although population change can help drive economic development, some countries could become victims of their own success if rising affluence brings challenges for energy and water consumption.

46 In 2019, four countries (Syria, Afghanistan, South Sudan and Myanmar) produced the great majority of the world's refugee population. Syria was the origin of 6.6 million refugees, more than twice that of the second-largest, Afghanistan (2.7 million).

47 Positive feedback loops may also cause serious environmental problems as they can cause resource depletion. For example, although the use of fossil fuel energy has increased our capacity to produce food, climate scientists say that it is creating a rise in global temperatures and a shift in rainfall patterns (due to the emission of greenhouse gases), which might negatively impact on food production from much of the world's existing farmland.

48 By 2025, the population of India is expected to surpass that of China. Thereafter, India's population is projected to continue growing for several decades to 1.5 billion in 2030 and 1.7 billion in 2050, while the population of China is expected to remain fairly constant at 1.4 billion until the 2030s, after which it is expected to slightly decrease. Nigeria's population is growing the most rapidly. Consequently, Nigeria is projected to have the third-largest population in the world by 2050.

49 Oil: over recent decades, various oil exploration sites around the world illustrate resource frontiers – Dubai and Abu Dhabi, Alaska North Slope, the east coast of Scotland, and Kashagan in the Caspian Sea.

50 For example, in the UK, mineral extraction such as quarrying and open cast mining must have an EIA carried out if the development is likely to have a significant impact on the environment by virtue of its nature, size or location. However, these three categories allow for considerable uncertainty about the need for an EIA, as the scale of impact will depend on whether it crosses a threshold that is not always clear. In simple terms, there is a three-stage process:
- Is the proposed development within a valid category of possible damaging activity?
- If so, either: (a) does it exceed the threshold set out for that category (often determined by local authorities)? or (b) is it in a 'sensitive area' such as an SSSI, national park, AONB etc?
- If so, is it likely to have a significant effect on the environment by virtue of its nature, size or location – such as aesthetics, air pollution, noise pollution or water pollution?

If the answer to all three of these questions is 'yes' then an EIA is required.

51 China and the USA produce the most energy in the world.

52 Biofuels production is concentrated in the USA, Brazil, Indonesia and Argentina.

53 TNCs help determine the price of a mineral ore as they:
- research for alternative materials
- invest (or fail to invest) in new areas of production
- apply technology to reduce costs of production and transport
- take note, or otherwise, of environmental concerns, which may include pressure to recycle

54 Water use has had severe environmental impacts. Some 50% of the world's wetlands have been drained, some 20% of freshwater fish have become extinct or are endangered, and the world's fourth largest lake, the Aral Sea, has shrunk to less than a third of its area due to commercial cotton irrigation.

55 Two miles north of its Cambodian border, along the Mekong River, Laos intends to construct the Don Sahong dam. The associated power project will affect water supplies and fishing in Cambodia, Vietnam and Thailand. Hence these three countries are demanding a say in the project.

56 In London, it is reported that over 90% of used water is recycled after intensive treatment at sewage works.

57 In Iceland, it has enabled heat to be provided to homes, open-air swimming pools and greenhouses. Due to such heat, Iceland is the only European country that is self-sufficient in bananas – grown in heated greenhouses.

58 Currently about 20% of the UK's electricity is generated by nuclear power stations, but this will decrease as nuclear plants begin to close as they became obsolete or need renewing. The government has plans to build new nuclear plants on old sites to fill the energy gap, such as Hinkley Point and Sizewell. New plants at both are being built by French and Chinese TNCs. However, the Japanese TNC Hitachi pulled out of a contract to build a new plant at Wylfa in North Wales.

59 Spain has substantial solar power capacity and its largest development at Olmedilla has a generating capacity of 50 MW, providing electricity for 40,000 homes. It is suggested that this plant will displace over 2 million tonnes of carbon dioxide emissions during its life span.

60 London: the congestion charge which encourages the use of public transport; the Greater London Low Emission Zone which makes the most polluting vehicles pay a daily charge when they drive anywhere in the area.

61 Iron and steel: steel's desirable properties and its relatively low cost make it the main structural metal in engineering and building projects around the world, accounting for about 90% of all metal used each year. 60% of iron and steel products are used in transportation and construction, 20% in machinery manufacture, and most of the remainder in cans and containers, and in various appliances and other equipment.

62 The resource potential of the Antarctic is already being exploited through biological prospecting. This involves taking Antarctic life forms and converting them for commercial use, for example using Antarctic green algae in skin cosmetics. This is approved activity, but biological prospecting offers insights into the tensions that might arise if resources such as oil, gas, iron ore, zinc, lead and coal were exploited.

Index

Note: page numbers in **bold** indicate location of key term definitions.